The Scientific Legacy of Fred Hoyle

Fred Hoyle was a remarkable scientist, and made an immense contribution to solving many important problems in astronomy. Several of his obituaries commented that he had made more influence on the course of astrophysics and cosmology in the second half of the twentieth century than any other person. This book is based on a meeting that was held in recognition of his work, and contains chapters by many of Hoyle's scientific collaborators. Each chapter reviews an aspect of Fred Hoyle's work; many of the subjects he tackled are still areas of hot debate and active research. The chapters are not confined to the discoveries of Hoyle's own time, but also discuss up-to-date research that has grown out of his pioneering work, particularly on the interstellar medium and star formation, the structure of stars, nucleosynthesis, gravitational dynamics, and cosmology. This wide-ranging overview will be valuable to established researchers in astrophysics and cosmology, and also to professional historians of science.

DOUGLAS GOUGH is the Professor of Theoretical Astrophysics, and Director of the Institute of Astronomy, at the University of Cambridge. He is an honorary professor at the University of London, an adjunct fellow at the University of Colorado, and a visiting professor of physics at Stanford University. His main research interest is the internal dynamics of stars.

Sir Fred Hoyle 1915–2001

The Scientific Legacy of Fred Hoyle

Edited by

DOUGLAS GOUGH
Institute of Astronomy, University of Cambridge

CAMBRIDGE
UNIVERSITY PRESS

CAMBRIDGE UNIVERSITY PRESS
Cambridge, New York, Melbourne, Madrid, Cape Town,
Singapore, São Paulo, Delhi, Tokyo, Mexico City

Cambridge University Press
The Edinburgh Building, Cambridge CB2 8RU, UK

Published in the United States of America by Cambridge University Press, New York

www.cambridge.org
Information on this title: www.cambridge.org/9781107402867

First published 2005
First paperback edition 2011

A catalogue record for this publication is available from the British Library

Library of Congress Cataloguing in Publication data

The scientific legacy of Fred Hoyle / edited by Douglas Gough.
     p.   cm.
   Includes bibliographical references and index.
   ISBN 0 521 82448 6
   1. Hoyle, Fred, Sir – Contributions in astrophysics.   2. Astronomy – History –
   20th century.   1. Gough, D. O.

   QB36.H75S35 2004
   523.01–dc22   2004045676

ISBN 978-0-521-82448-4 Hardback
ISBN 978-1-107-40286-7 Paperback

Contents

Contributors

David Arnett
Steward Observatory
University of Arizona
Tucson AZ 85721
USA

John D. Barrow
Department of Applied Mathematics and Theoretical Physics
Centre for Mathematical Sciences
University of Cambridge
Wilberforce Road
Cambridge
CB3 0WA
UK

Sir Hermann Bondi
Churchill College
University of Cambridge
Storey's Way
Cambridge
CB3 0DS
UK

E. Margaret Burbidge
Center for Astrophysics and Space Sciences
University of California at San Diego
9500 Gilman Drive
La Jolla CA 92093
USA

Geoffrey Burbidge
Center for Astrophysics and Space Sciences
University of California at San Diego

9500 Gilman Drive
La Jolla CA 92093
USA

George Efstathiou
Institute of Astronomy
University of Cambridge
Madingley Road
Cambridge
CB3 0HA
UK

John Faulkner
University of California/Lick Observatory
Department of Astronomy and Astrophysics
University of California Santa Cruz
Santa Cruz CA 95064
USA

Malcolm S. Longair
Cavendish Astrophysics Group
Cavendish Laboratory
University of Cambridge
Madingley Road
Cambridge
CB3 0HE
UK

Jayant V. Narlikar
Inter-University Centre for Astronomy and Astrophysics
Pune 411 007
India

Martin Rees
Institute of Astronomy
University of Cambridge
Madingley Road
Cambridge
CB3 0HA
UK

Wallace L. W. Sargent
Astronomy Department
California Institute of Technology
Pasadena CA 91125
USA

Philip M. Solomon
Department of Physics and Astronomy
Stony Brook, State University of New York
Stony Brook NY 11794
USA

Chandra Wickramasinghe
Cardiff Centre for Astrobiology
Cardiff University
2 North Road
Cardiff
CF10 3DY
UK

Foreword

MARTIN REES

Fred Hoyle's varied and prolific output spanned more than 60 years. Indeed, throughout the entire period 1945–70 he was preeminent among astrophysicists in the range and influence of his contributions. This one-day memorial meeting focused on his research contributions, but of course all those who knew him – and, indeed, the wide public – admired him for other reasons too. His engaging wit and relish for controversy, retained throughout his long life, gained him a high public profile. He had a wide following as a popularizer of science and as a successful writer of science fiction. He also played an active organizational role in UK science. Fred died on 20 August 2001 in Bournemouth, England. He was physically and mentally robust until the year before his death, during which he suffered a series of strokes.

Born on 24 June 1915 in Bingley, Yorkshire, he was the son of a wool merchant. He attended the local grammar school, from which he gained a scholarship to Emmanuel College, Cambridge, where he read mathematics. He graduated with a BA in mathematics in 1936, winning the Mayhew Prize for his outstanding performance. He was elected to a fellowship at St John's College, Cambridge, in 1939 for work on beta decay. His shift towards astrophysics was stimulated by his colleague Raymond Lyttleton, with whom he wrote papers on accretion and stellar evolution. Although accretion plays such a great role, few of the younger generation of astrophysicists will have read these papers. I certainly was unaware of the remarkably detailed discussion of cooling processes involving molecular hydrogen in a 1940 paper by Hoyle and Lyttleton, described in Phil Solomon's contribution.

During the years of the Second World War, Fred was engaged mainly on technical problems related to radar. He found himself working with Hermann Bondi and Thomas Gold; in spare moments the trio discussed astronomy. The most celebrated outcome of this collaboration was the steady-state cosmology, put forward in two papers in 1948. Bondi and Gold's arguments were general (almost philosophical). But the Hoyle model was more specific: he introduced a negative-pressure C-field into Einstein's equations. As Fred enjoyed pointing out, this formulation was, in some sense, a precursor of currently fashionable inflationary models. The steady-state theory was a serious contender for 15 years. (Its three promoters were articulate and effective advocates. However, it is fair to say that their voices did not carry across the Atlantic, where the theory never acquired the visibility it had in the UK). It had the virtue of being testable, and was the focus of often acrimonious controversy, especially with Martin Ryle and other radio astronomers. Fred held out against Big-Bang theory, even post-1965, when the discovery of the microwave background led most cosmologists to favour it. He nonetheless contributed important studies of Big-Bang nucleosynthesis with Roger Tayler, Willy Fowler, and Bob Wagoner.

From 1945, Hoyle was based in Cambridge, first as lecturer in mathematics, and subsequently, from 1958, as the Plumian Professor of Astronomy. But he derived stimulus from frequent visits to the USA. He was one of the early users of Cambridge's EDSAC computer for his studies of red giants and stellar evolution. At Princeton University, he and Martin Schwarzschild modelled the evolution of low-mass stars right through to the red-giant branch. He spent much time at Caltech, where he followed up the ideas adumbrated in his famous 1946 paper, 'The synthesis of the elements from hydrogen', in a long and fruitful collaboration with Willy Fowler on nuclear processes in stars and supernovae. This research was codified in a classic 1957 article, universally referred to as 'B²FH' (published in *Reviews of Modern Physics*), which he and Fowler coauthored with Geoffrey and Margaret Burbidge. Many of us felt that Fred should have shared Fowler's 1983 Nobel Prize in Physics, but the Royal Swedish Academy of Sciences later made partial amends by awarding him, with Edwin Salpeter, its 1997 Crafoord Prize.

Throughout the 1950s and 1960s Fred kept up his wide-ranging interests in solar physics, the origin of the Solar System, the structure of galaxies, and the nature of gravity. The discovery of quasars in the 1960s led to a stream of stimulating papers, many coauthored with the Burbidges, on supermassive objects and various aspects of high-energy astrophysics. Fred was at the leading edge of all these developments. As Malcolm Longair describes in this book, those who attended his lectures had the privilege of following the development of his theories in 'real time'.

Committee work and administration held little attraction for Fred. Nonetheless, especially during the 1960s and early 1970s, he served effectively on the Science Research Council and the Council of the Royal Society, among other UK bodies. In Cambridge, his energetic advocacy and fundraising led to the creation, in 1966, of the Institute of Theoretical Astronomy. Its main building, now named after him, was modelled on the University of California's Institute of Geophysics and Planetary Physics in La Jolla, California, although it overlooks a field of cows rather than the Pacific Ocean. He had a remarkably 'hands on' role in the detailed design, and all the occupants of this building have cause to be grateful to him. (He believed that wide corridors were not a waste of space but would encourage an interactive atmosphere. He even convinced the budgeters that it was 'economical' to carpet the building throughout, since otherwise acoustic cladding would be needed, which would be more expensive.) It was specially timely that this memorial meeting was the first conference to be held in the Hoyle Building after its renovation and extension – it is now substantially enlarged, but in a style which fully maintains the environmental qualities that Fred valued and created. The Institute of Theoretical Astronomy opened in 1967 and quickly made an international mark. Key ideas on supernovae and explosive nucleosynthesis were developed by Willy Fowler, David Arnett, Don Clayton, Stan Woosley and other colleagues during their regular summer visits.

Fred's regular collaborators Jayant Narlikar and Chandra Wickramasinghe were part of the Institute's full-time staff. In addition, a lively group of postdoctoral scientists benefited from the stimulating environment of the Institute. I was privileged to be one of these, along with Brandon Carter, Stephen Hawking, Joe Silk and many others. A lot of our research was orthogonal (or even contradictory) to his own. At a Vatican 'Study Week' in 1970, my talk, given immediately after his, was quite out of line with his interpretation of radio source counts, but scientific disagreements did nothing to diminish his friendly support. On the broader UK scene, Hoyle's role was pivotal in establishing the Anglo-Australian Observatory, in the early 1970s. As a result, for the first time, UK astronomers had guaranteed access to a world-class optical telescope.

The dispute that led to Fred's premature retirement from Cambridge in 1972 was deeply regrettable. He thereafter based himself for many years in a remote part of the Lake District (hill-walking being one of his lifelong enthusiasms) before moving to the more sedate environs of Bournemouth. His consequent isolation from the broad academic community was probably detrimental to his own science; it was certainly a sad deprivation for the rest of us. But there was no let-up in his productivity. His scientific writings continued throughout the 1980s and 1990s, and dealt, often controversially, with topics as disparate as Stonehenge, panspermia, Darwinism, palæontology, and viruses from space. It was perhaps unfortunate that he gained more public attention for these excursions than he ever had for his work on nucleosynthesis. But he never lost his focus on cosmology, nor the hope of achieving a deeper synthesis: his book *A Different Approach to Cosmology: From a Static Universe through the Big Bang towards Reality*, coauthored with G. Burbidge and Narlikar, appeared in 2000.

His lifelong success as a popularizer started in 1950 – in the pre-Sagan era, long before the dominance of television – with a celebrated series of radio talks. Huge numbers of people (including many who later achieved scientific distinction) were inspired by these talks, by books such as *Frontiers of Astronomy*, and by his lectures. Throughout his life, he retained the distinctive accent of his native Yorkshire.

Fred's first novel, *The Black Cloud* (Harper 1957), about an alien intelligence embodied in a cloud of interstellar gas, has achieved classic status. It was followed by a dozen others, including *A for Andromeda*, coauthored with John Elliot, which was dramatized as a television series, *Ossian's Ride* (1959) and *October the First is Too Late* (1966). Some of his books, including those he wrote for children during his later years, were coauthored with his son Geoff. His autobiography *Home Is Where the Wind Blows: Chapters from a Cosmologist's Life* (University Science Books 1994) sensitively evokes his early life in Yorkshire and offers entertaining perspectives on later academic disputes.

The memorial meeting in Cambridge offered a chance to review and celebrate Fred's enduring insights into stars, nucleosynthesis, and the large-scale universe. Some of these rank among the great achievements of twentieth-century astrophysics. Most of us were aware of his 'highlight' contributions, but (along, probably with most of the audience) I learnt about diverse aspects of his work which were completely new to me. His theories were unfailingly stimulating, even when they proved transient. He will be remembered with fond gratitude not only by colleagues and students, but by a much wider community who knew him through his talks and writings.

Preface

Fred Hoyle was one of the great figures in twentieth-century theoretical astrophysics. His many scientific writings on an extremely wide range of astronomical subjects, from solar and stellar physics to cosmology and panspermia, bear witness to his penetrating mind and his remarkable versatility. In addition, he captured the imagination of the public with his popular books, his radio broadcasts and his public lectures. A great deal of what Fred did was controversial at the time, but is becoming less so as the years go by, as generally accepted scientific wisdom moves more and more towards the essence of his ideas, if not necessarily the details of his application of them to astronomy: an example, discussed in this book, is the introduction of his C-field into the dynamical equations describing the evolution of the Universe, a negative-energy field which has subsequently reappeared in the currently almost universally accepted inflationary scenario. Yet much of Fred's work was quite obviously sound from the outset, such as his pioneering monumental study, with Willy Fowler and Geoff and Margaret Burbidge, of the creation of the chemical elements, and his seminal work with Martin Schwarzschild on the evolution of low-mass stars and on the structure of red giants.

For three decades of his scientifically most productive life, Fred was based in Cambridge. He was a Fellow of St John's College from 1939 and, after intermission for the Second World War, became a lecturer in applied mathematics in the Department of Applied Mathematics and Theoretical Physics in 1945, and subsequently Plumian Professor of Astronomy and Experimental Philosophy. In 1966 Fred founded the Institute of Theoretical Astronomy (which subsequently amalgamated with the University Observatories to form the Institute of Astronomy), and was its Director until his departure from Cambridge in 1972. It was fitting, therefore, that an international meeting to honour his scientific life should take place in Cambridge. The meeting was held on 16 April 2002, at the Institute of Astronomy and St John's College. It was attended by Fred's friends and collaborators, by younger scientists some of whom had not known Fred personally, and also by many of Fred's family. It was not merely retrospective, but looked forward to new advances. This book is a record of the proceedings of that meeting. It contains discussions not only of work carried out by Fred and his collaborators in the context of present-day thinking, but also of up-to-date developments in astronomy and cosmology that have grown out of Fred's ideas.

I thank all those who have helped with the production of this book, particularly Bob Carswell, who wrote the first chapter with the help of his video recording of Wal Sargent's talk, to Di and Richard Sword who polished the LaTeX and the diagrams, to Mark Hurn who assembled the index, and to Jacqueline Garget of Cambridge University Press whose patience was truly tried by delays in receiving the final text.

1

Fred Hoyle's major work in the context of astronomy and astrophysics today

WALLACE L. W. SARGENT

Astronomy Department, California Institute of Technology

1.1 Hoyle's major contributions

It was a great privilege to be asked to open the proceedings for the Hoyle science retrospective. I went to the Institute in the second year of its existence in 1968, and then for five consecutive years until Fred resigned. Since then I have gone less often, but it is always a great pleasure to do so. In many ways Fred could have no more fitting memorial than this Institute, which continued to grow in stature even after Fred resigned, and I am sure it will continue to do so.

The organizers asked me as a former close friend and colleague of Fred, but one who is an observing astronomer, to summarize work which is largely theoretical. I presume that this is because they wanted a broad-brush overview of the main themes of Fred Hoyle's research. These are:

- Accretion 1941–47
- Stellar structure and evolution 1942–64
- Nucleosynthesis 1946–74
- Cosmology 1948–2001
- Interstellar dust 1962–2001

There is also another thread to his research, panspermia, etc. in the years 1974–2001, which is less directly related to his time in Cambridge, and which I shall not describe here.

The Scientific Legacy of Fred Hoyle, ed. D. Gough.
Published by Cambridge University Press. © Cambridge University Press 2004.

1.2   Accretion theory

The first theme in Fred's work was accretion, with a number of studies published in the 1940s. Hoyle and Lyttleton (1941) and Bondi and Hoyle (1944) suggested that accretion of interstellar gas would increase the masses of stars significantly during their lifetimes. Also there was the suggestion that the solar corona is the result of accretion, resulting from convective dissipation (Bondi, Hoyle and Lyttleton 1947). While accretion by normal stars is no longer thought to be very important, some stars do continue to accrete matter after they are formed. In that sense accretion is still an important topic in stellar physics, and, modified by ambipolar diffusion of magnetic fields, is now thought to be important in the late stages of star formation.

Accretion is also important in considering how the first massive stars in the Universe formed at redshifts $z \sim$ 6–15. These stars produced the first heavy elements, a theme in which Fred was very interested in his later work. According to computer simulations it appears that these stars first formed as a nucleus and then grew by intergalactic gas falling in on the seeds.

1.3   Stellar structure and evolution

One of Fred's major themes for several years was stellar structure and evolution. His first work in this area was on the structure of red giants (Hoyle and Lyttleton 1942). In 1945, in the days long before global relaxation methods were introduced into astrophysics, he introduced a new method for solving the equations determining the structure of a star with a convective core. Rather than integrating from the centre outwards, as was the practice of the time, and thereby suffer having to deal with the extreme sensitivity of the solution near the surface to variations in the initially unknown conditions at the centre, Fred integrated inwards appropriately sealed equations that depend on only a single parameter and are not so stiff, the parameter being determined iteratively by matching onto a solution of the Lane–Emden equation representing the convective core (Hoyle 1945).

One of the great accomplishments for which Fred is still renowned is the theory of nucleosynthesis in stars. It was a prediction of the rate of the triple-α reaction to form ^{12}C, which had earlier been worked on by Salpeter. A modification was introduced by Hoyle in 1953, which involved predicting the existence of an excited state in the ^{12}C which resulted in a higher rate for the $3\,^4$He $\rightarrow\,^{12}$C reaction, and a slow-down for ^{12}C $+\,^4$He $\rightarrow\,^{16}$O. This is explained in more detail by Dave Arnett in these proceedings. This enabled the Universe to save itself from becoming mostly oxygen, and kept a significant amount of carbon to produce life.

Later Hoyle and Schwarzschild (1955) calculated the evolution of Population II stars from the main sequence to red giants. These stars are found in globular clusters, and are thought to be the oldest stars in the Galaxy. The result of these calculations was the first accurate measurement (or, perhaps, statement) of the age of the oldest stars in the Galaxy. The background to this work, the work itself and its repercussions are discussed in considerable detail in these proceedings by John Faulkner.

At about that time Fred Hoyle returned to England and began to use digital computers to calculate stellar structure (Haselgrove and Hoyle 1956, 1958). Those who have read *The Black Cloud* will remember that the progress of the black cloud was calculated on a computer using machine language. This is a difficult task, and the description is based on the personal experience Fred had in calculating stellar interiors.

As a result of their calculations, Fred and his collaborators produced the standard picture of red giants: these stars have isothermal, inert, helium cores, thin hydrogen-burning shells, and extended convective envelopes. They also arrived at realistic ages, of about 10 billion years, for the oldest Galactic stars.

Later, Fred collaborated with W. Fowler, and they were the first to note the roles of Type I and Type II supernovae in making heavy elements. They correctly surmised that Type I supernovae arise from the explosion of degenerate matter, of the type found in white dwarfs. We now believe that such supernovae occur in binary systems. Type II supernovae arise from the implosion and subsequent explosion of non-degenerate stellar cores (Hoyle and Fowler 1960; Fowler and Hoyle 1964). This remains the standard picture of supernova explosions today, with the exception that in those early days the role of neutrino transport was not realized, and the sites of the e-process and the r-process were incorrectly assigned.

1.4   Nucleosynthesis in stars

The work that I personally admire most in all of Fred's many achievements is that on nucleosynthesis in stars.

By the mid 1940s heavy elements were thought to originate in an initial dense hot phase of the Universe, and it was Fred who first realized that stars can produce heavy elements and that these can be spread into the surrounding interstellar medium by explosive processes or by stellar winds. He was also the first to realize that in massive stars which evolve to have very hot dense interiors statistical equilibrium would produce the iron-peak elements (later dubbed the 'e-process'). This, followed by explosive ejection, would enrich the interstellar gas in these elements (Hoyle 1946). This work focused people's attention on the idea that all heavy elements are made from hydrogen by nucleosynthesis in stars. This is the standard paradigm today, except for D, ^{4}He, ^{3}He, ^{7}Li, most of which is produced in the hot Big Bang.

This work was followed by the suggestion (Hoyle 1954) that the synthesis of carbon to nickel is due to successive thermonuclear buildup from hydrogen in hotter and hotter stars.

In 1957 Burbidge, Burbidge, Fowler and Hoyle wrote an amazingly prescient review article (Burbidge *et al.* 1957) on the general question of the abundances of the elements. This review contained a large amount of original work, and a systematic discussion of the many processes involved. These were the alpha process (helium capture); the e-process described above; the r-process, which is the addition of neutrons to iron-peak elements on a rapid timescale; the s-process, which is the same on a slower timescale; the p-process for the addition of protons to nuclei; and the x-processes for the production of the light elements Li, Be, B. This landmark paper came to be known as 'B^{2}FH'.

When I first went to Caltech in 1959 to work on the abundances of the elements, almost every talk and seminar in the subject began 'According to B^{2}FH...'. The speaker might then go on to say that something in B^{2}FH was wrong, but usually the conclusion was that for the aspect they were considering B^{2}FH was correct.

B^{2}FH still provides the framework for present-day discussions of cosmic element abundances. While there have been modifications to the ideas that were put forward then, there have been no revolutionary different descriptions. The question of the site(s) of the r-process is a very active current field, particularly for the oldest stars. For an extended review of the situation 40 years after B^{2}FH see Wallerstein *et al.* (1997).

Somewhat later, Fowler and Hoyle (1960) began the science of nuclear cosmochronology, which extends to the Cosmos the ideas that had been used in geochronology to age-date systems. Using particularly the long-lived isotopes of Th and U, they inferred an age for the Universe of 11 billion years.

Hoyle and Fowler (1960) also studied nucleosynthesis in supernovae. One of their innovations was the realization that, at very high temperatures of $\sim 10^9$ K, pair production of neutrinos and antineutrinos, and positrons and electrons would cause a massive star to become unstable (Fowler and Hoyle 1964). This is particularly important for the study of the first stars, which must have zero metallicity. These stars are now thought to be massive, and could in some cases destroy themselves completely and leave no remnant as a result of this 'pair instability'.

1.5   Cosmology

Fred's contribution to cosmology is what he was best known for by the general public. In 1948 he, and Bondi and Gold, in two separate papers (Hoyle 1948; Bondi and Gold 1948), put forward the idea that the Universe is in a steady state. Fred's contribution was the introduction of an extra term $C_{\mu\nu}$ into the Einstein field equations. This extra term represents the creation of matter. At least initially the form in which matter was created was not specified, although of course it had to be electrically neutral. More recent theory, which explains the isotropy and homogeneity of the Universe, namely the 'inflation' theory, has a metric which is identical to that of the C-field cosmology.

During 1955–65 there was the controversy over the radio-source counts. This topic is covered more fully in Malcolm Longair's contribution to this volume. In fact, there is very little written by Fred about the radio-source counts having a power law slope steeper than the -1.5, which is the Euclidean value. However, Fred did suggest that the angular-diameter vs. redshift relation for radio sources could be used to distinguish the Steady State from the evolving Einstein–de Sitter cosmologies without appealing to source counts (Hoyle 1959). Moreover, he did point out in papers with Narlikar (Hoyle and Narlikar 1961, 1962) that a modification of the Steady-State theory could give $\log N$–$\log S$ slopes steeper than -1.5, and so the theory survived the requirement of observations.

The idea that massive objects and relativistic objects in galactic nuclei play a role in explaining the violent phenomena which were discovered through radio astronomy was introduced by Hoyle and Fowler (1963), and Hoyle, Fowler, Burbidge and Burbidge (1964). They did not involve black holes explicitly, but there was certainly the notion that objects in which general relativity is important are involved. This is a view that we now believe is correct.

A major contribution that Hoyle made to cosmology was on the production of the light elements. In a paper in *Nature*, Hoyle and Tayler (1964) suggested that all of the helium in the Universe could be produced in an early dense phase in the Universe or in massive stars. Later Wagoner, Fowler and Hoyle (1967) produced a detailed paper in which the synthesis of D, ^{4}He, ^{3}He, ^{7}Li was discussed, and found to be in accord with microwave background temperature. There was also the proviso that the same results could probably come from massive stars at high temperatures. This paper began the industry of calculating the cosmological density parameter from the abundances of the light elements. In a much later paper (Burbidge and Hoyle 1998) it was concluded that synthesis of the light elements in stars is possible.

Fred spent quite some time, particularly later in his career with Burbidge and Narlikar, modifying the Steady-State theory, because in some instances it was not in accord with observations. The three became interested in quasars, and suggested a 'local' hypothesis, in which quasars are ejected at relativistic speeds from nuclei of nearby galaxies (Hoyle and Burbidge 1966). Hoyle and Narlikar (1972) produced a conformally invariant gravitational theory in which the Universe has 'another side' prior to the Robertson–Walker singularity. Hoyle and his collaborators also considered modifications to the Steady-State theory that would be consistent with the extreme isotropy of the microwave background. One possibility proposed was that the isotropy is the result of thermalized stellar radiation from the dense epoch as the Universe passed from the 'other side', and could be linked to the cosmic helium abundance (Hoyle 1975). Another, in 1980, was a revision of the theory in which new galaxies are generated in a series of small bangs. One interesting modification was a quasi-Steady State, with a major creation event when the Universe had a mean density of 10^{-27} g cm^{-3} (Hoyle, Burbidge and Narlikar 1993).

Hoyle and collaborators struggled for several years to explain the general isotropy of the microwave background and the small fluctuations on large angular scales discovered by COBE in terms of the Steady State or its modifications. All involved thermalizing starlight using ingenious mechanisms (e.g. iron or graphite whiskers), but none were really successful.

1.6   Interstellar dust

Interstellar dust may sound less grandiose a topic than the nature of the Universe and the origin of the elements, but it is very important for several reasons. For example, it appears that dust can be formed in the atmospheres of even the earliest stars, and that light from galaxies even at the earliest times in the Universe suffers from dust extinction. This makes it very hard to obtain a complete census of galaxies during the first phases of their evolution.

The first piece of work in this area was by Hoyle and Wickramasinghe (1962), who suggested that graphite particles are an important constituent of dust grains. This was then followed by a paper pointing out the importance of graphite–ice grains for ultraviolet extinction (Hoyle and Wickramasinghe 1963). Together with Don Clayton, Hoyle and Wickramasinghe made, in 1975, the first suggestion that the interstellar particles, which we can study physically when they are found in meteorites, actually have their origin in the atmospheres of novae and supernovae. Before that it had been thought that dust grains were assembled in interstellar space.

Fred was a pioneer in the early use of infrared spectra to investigate dust properties. He was one of the first to realize that some of the infrared spectral features could be due to organic compounds as well as the silicates and water, which had already been identified (Wickramasinghe, Hoyle and Nandy 1977). Further work involved polysaccharides (hydrocarbons) in grains (Hoyle and Wickramasinghe 1977), and the realization that the 2200A feature in the infrared spectra is not due just to graphite particles (Wickramasinghe, Hoyle and Nandy 1977).

Fred's work on dust is also relevant in a more recent context because at high redshifts we can determine the abundances of heavy elements in the interstellar matter in very distant galaxies. However, these abundances are modified by the degree to which each of these separate elements is taken up into dust grains and so taken out of the gas phase. So considerations about the formation and chemistry of dust grains that were initiated by Hoyle and Wickramasinghe all those years ago remain important today.

1.7   Concluding remarks

My first knowledge of Fred was on the radio when I came home from the Scunthorpe Technical High School one evening in February 1950. Then I heard Fred first broadcast in what became a series of six talks on the nature of the Universe. It was at that time that I realized that even people from Scunthorpe with accents like mine could do this kind of work, a realization for which I shall forever be grateful to Fred.

References

BONDI, H. & GOLD, T. 1948 *MNRAS*, **108**, 252

BONDI, H. & HOYLE, F. 1944 *MNRAS*, **104**, 273

BONDI, H., HOYLE, F. & LYTTLETON, R. A. 1947 *MNRAS*, **107**, 184

BURBIDGE, G. R. & HOYLE, F. 1998 *ApJ*, **509**, 1

BURBIDGE, E. M., BURBIDGE, G. R., FOWLER, W. A. & HOYLE, F. 1957 *Rev. Mod. Phys.*, **29**, 547

CLAYTON, D. D., HOYLE, F. & WICKRAMASINGHE, N. C. 1975 *BAAS*, **7**, 553

FOWLER, W. A. & HOYLE, F. 1960 *AJ*, **65**, 345

   1964 *ApJS*, **9**, 201

HASELGROVE, C. B. & HOYLE, F. 1956 *MNRAS*, **116**, 515

   1958 *MNRAS*, **118**, 519

HOYLE, F. 1945 *MNRAS*, **105**, 23

   1946 *MNRAS*, **106**, 343

   1948 *MNRAS*, **108**, 372

   1954 *ApJS*, **1**, 121

   1959 *IAU Symposium no. 9*, ed. R. N. Bracewell (Stanford: Stanford University Press) p. 529

   1975 *ApJ*, **196**, 661

HOYLE, F. & BURBIDGE, G. R. 1966 ApJ, **144**, 534

HOYLE, F. & FOWLER, W. A. 1960 ApJ, **132**, 565

1963 MNRAS, **125**, 169

HOYLE, F. & LYTTLETON, R. A. 1941 MNRAS, **101**, 225

1942 MNRAS, **102**, 218

HOYLE, F. & NARLIKAR, J. V. 1961 MNRAS, **123**, 133

1962 MNRAS, **125**, 13

1972 MNRAS, **155**, 305

HOYLE, F. & SCHWARZSCHILD, M. 1955 ApJS, **2**, 1

HOYLE, F. & TAYLER, R. J. 1964 Nature, **203**, 1108

HOYLE, F. & WICKRAMASINGHE, N. C. 1962 MNRAS, **124**, 417

1963 MNRAS, **126**, 401

1977 MNRAS, **181**, 51

HOYLE, F., BURBIDGE, G. R. & NARLIKAR, J. V. 1993 ApJ, **410**, 437

HOYLE, F., FOWLER, W. A., BURBIDGE, G. R. & BURBIDGE, E. M. 1964 ApJ, **139**, 909

WAGONER, R. V., FOWLER, W. A. & HOYLE, F. 1967 ApJ, **148**, 3

WALLERSTEIN, G. et al. 1997 Rev. Mod. Phys., **69**, 995

WICKRAMASINGHE, N. C., HOYLE, F. & NANDY, K. 1977 Ap Space Sci., **47**, L9

2

Sir Fred Hoyle and the theory of the synthesis of the elements

DAVID ARNETT
Steward Observatory, University of Arizona

Some of Fred Hoyle's pioneering ideas about the site and the nature of the synthesis of the elements are examined in a modern context of theory, experiment and observations. Hoyle's ideas concerning the nucleosynthesis cycle of stellar birth and death, rotational instability of supernovae, the onion-skin model of presupernovae, neutronization, nuclear statistical equilibrium and core collapse, thermonuclear supernovae, nucleosynthesis processes and freeze-out are discussed. The history of the clash of theory and experiment on the second excited state of ^{8}Be and helium ignition in red giants is reviewed.

2.1   Introduction

Sir Fred Hoyle (1915–2001) was the architect of the theory that the naturally occurring nuclei were synthesized from hydrogen by thermonuclear burning in stars, and especially in supernova explosions (Hoyle 1945). Today we would modify this slightly to include some primordial ^{4}He along with traces of deuterium, ^{3}He, and ^{7}Li from the Big Bang (Wagoner, Fowler and Hoyle 1967) as the original fuel for synthesizing the rest of the nuclei. Many of his ideas were already contained in two early papers, Hoyle (1946) and Hoyle (1954), which preceded the famous paper by Burbidge, Burbidge, Fowler and Hoyle (1957) and later work with W. A. Fowler (especially see Hoyle and Fowler 1960 and Fowler and Hoyle 1964).

The Scientific Legacy of Fred Hoyle, ed. D. Gough.
Published by Cambridge University Press. © Cambridge University Press 2004.

It is now possible to test these ideas: by direct experiment, by improved observation and by numerical simulation. We shall examine a few of the most spectacular examples.

2.2   The stellar cycle of nucleosynthesis

Work on accretion of interstellar gas by stars (Hoyle and Lyttleton 1941 and Bondi and Hoyle 1944) led to a study of how the properties of such material could lead to its condensation into stars in a galactic context (Hoyle 1945). The idea that stars were formed by accretion of interstellar gas in turn led to the idea that the subsequent generations of stars and their nucleosynthesis was an ongoing process (a 'nucleosynthesis cycle') in the evolution of galaxies (Hoyle 1946), unlike the 'one shot does it all' ideas of nucleosynthesis then common (e.g. Chandrasekhar and Henrich 1942 and Alpher and Herman 1953).

Figure 2.1 captures this nicely. The small ringed object near the centre of the image is the remnant of Supernova 1987A in the Large Magellanic Cloud.

Figure 2.1 Star formation and Supernova 1987A in the Large Magellanic Cloud (Hubble Heritage image: NASA/STScI).

Figure 2.2 Star formation and dust clouds in the Whirlpool Galaxy (Hubble Heritage image: NASA/STScI).

The glowing gas is illuminated by the ionizing radiation from a generation of young massive stars which have just formed from interstellar gas and not yet evolved to become supernovae or white dwarfs. The gas, which is still in the form of a molecular cloud, obscures background gas, making the apparent 'holes' of darkness. SN1987A was observed to eject significant amounts of matter composed of elements heavier than H and He; its light was powered by the decay of ^{56}Co, which itself decayed from ^{56}Ni freshly synthesized in the explosion.

Figure 2.2 shows this on the galactic scale, in the Whirlpool Galaxy. The dark dust lanes are the sites of relatively dense molecular clouds, in which bursts of star formation are occurring. The bright regions are caused by young massive luminous OB stars and the gas they ionize. Nucleosynthesis was not just a narrowly defined subject that provided energy sources for stellar evolution and spice for

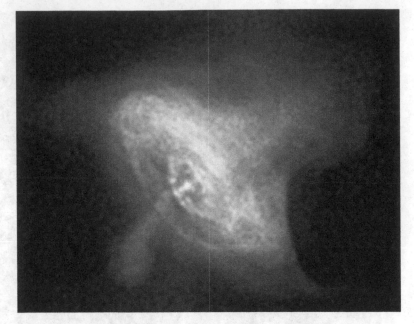

Figure 2.3 Crab Nebula remnant and pulsar in X-rays (Chandra image: NASA/CXC/SAO).

a nuclear physics lecture – although it served those functions well – but a fundamental part of the grand scheme of nature: the evolution of galactic systems.

2.3   Rotational instability and core collapse

Hoyle (1946) delineated the conditions of nuclear statistical equilibrium, argued that a key issue was the 'freeze-out' of nuclear reactions, suggested that this required an explosive process, and that this explosion was related to a rotational instability in the collapsing core. In an uncharacteristic fit of caution he assumed the core would become a massive white dwarf rather than the then speculative option of a neutron star or black hole. Later work with W. A. Fowler (Hoyle 1954, Fowler and Hoyle 1964) included the explosive fusion of ^{16}O as an energy source, but kept the rotational dynamics as a key feature.

Figure 2.3 is an image from the Chandra X-ray Observatory of the region surrounding the pulsar in the Crab Nebula, the remnant of the supernova of AD 1054 which was specifically discussed by Hoyle (1946). The mere shape of the image evokes a sense of rapid rotation; the pulsar itself rotates with a period of 0.033 seconds.

Figure 2.4 is an expanded image from the Hubble Space Telescope of the remnant of Supernova 1987A; the complex ring structure is compelling evidence

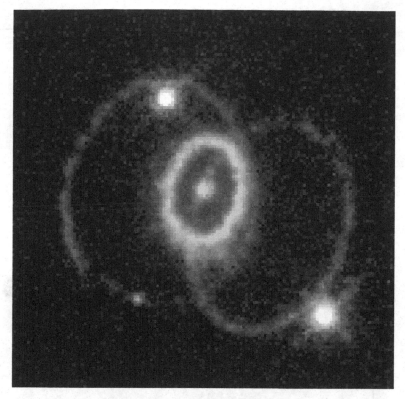

Figure 2.4 Ring structure around Supernova 1987A (NASA; C. Burrows (ESA/STScI and NASA)).

for the importance of rotation for at least shaping the material ejected prior to explosion (the rings), and perhaps for the instability itself.

In the intervening years since the Fowler and Hoyle (1964) discussion of the supernova mechanism, the idea of a spherically symmetric collapse and explosion had two irrestiable attributes: it seemed viable, and it was calculable with existing computers. This comfortable doctrine itself has now been exploded. Figure 2.5 is an infrared image from the Magellan telescope showing the host galaxy and a circle indicating Supernova 2001ke, whose spectrum was obtained with the Magellan 6.5-m telescope (Garnavich, et al. 2003) about 12 hours after the location of the gamma-ray burst GRB 011121 was determined by the BeppoSAX satellite (Piro 2001). The association of the unusual Type Ic Supernova 1998bw with GRB 980425 (e.g., Patat et al. 2001) and that of SN2001ke with GRB 011121 (Bloom et al. 2002, Price et al. 2002, and Garnavich et al. 2003) now give compelling evidence for the origin of at least some of the long duration gamma-ray bursts in the core collapse of massive stars. Synthesis of the observed light curves and spectra of SN1998bw requires an energy 7 to 50 times higher than the

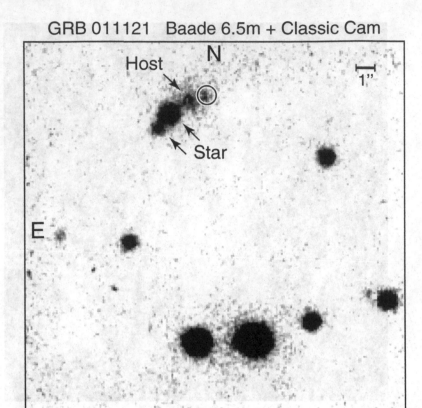

GRB 011121   Baade 6.5m + Classic Cam

UT 01112830   Jband 5.7 hours

Figure 2.5 The site of gamma-ray burst GRB 011121 and Supernova 2001ke in an infrared image from the Magellan 6.5-m Telescope (Garnavich *et al.* 2003, Magellan infrared image).

nominal value of 1.0×10^{51} ergs. These large energies suggest that the dominant source is gravitational rather than thermonuclear energy, closer to Hoyle (1946) than to Fowler and Hoyle (1964). Although the luminosities are high, indicating more than typical ^{56}Ni production (hence greater thermonuclear energy release), the high velocities demanded by the spectra indicate still higher kinetic energies.

The issue of the mechanism by which supernovae turn collapse into explosion is still unresolved, but it looks as if rotation may be a key aspect of an acceptable mechanism.

2.4   The 'onion-skin' model of presupernovae

The 'onion-skin' model of stars prior to supernova explosion was developed by Hoyle (1946). He showed that a presupernova would naturally develop

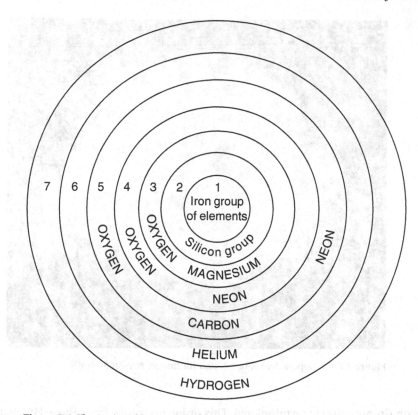

Figure 2.6 The 'onion-skin' model of a presupernova: Figure 47 from *Frontiers of Astronomy*, p. 211.

a layered structure, reflecting the sensitive temperature dependence of nuclear statistical equilibrium and of nuclear reactions. In the related context, Hoyle and Schwarzschild (1955) showed that the existence of red-giant stars depended upon stars being inhomogeneous, that is, poorly mixed.

Stars are so large that microscopic mixing takes a significant time relative to the rate of thermonuclear evolution. Thus the layering in temperature needed for hydrostatic equilibrium implies a layering in thermonuclear ashes. Convection, which is now known to be driven strongly by neutrino emission interacting with thermonuclear burning (see Arnett 1996 and references therein), does not change the qualitative picture. Upon seeing the presentation of some of the simulations of convection at the Crafoord Prize conference honouring him and E. E. Salpeter, Sir Fred's comment was 'Be careful, Dave. You're on a slippery slope.' He was right; we have much more work to do on this issue.

Figure 2.6 shows Hoyle's sketch of a presupernova from *Frontiers of Astronomy* (Hoyle 1955), in which he defined seven zones. These were separated by six burning stages, with the fuels being hydrogen, helium, carbon, neon, oxygen, and

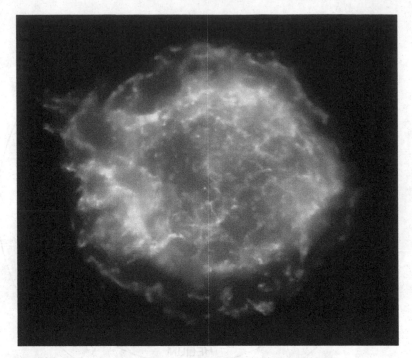

Figure 2.7 Cassiopeia A in X-rays (Chandra image: NASA/CXC/SAO).

silicon (in present-day terminology). This ordering comes from Hoyle (1954), and has been repeatedly confirmed by computer simulation (see Arnett 1996 and references therein, Arnett 1974, Woosley and Weaver 1981).

If compositional layering exists before the explosion, it is unlikely to be microscopically mixed during the short time of the explosion itself. While macroscopic instabilities, such as Rayleigh–Taylor and Richtmeyer–Meshkov, are important in realistic models of supernovae (Fryxell, Müller and Arnett 1991), they do not seem likely to give enough microscopic mixing to prevent young supernova remnants from bearing indications of their previous layering.

This seems to be the case, as Figure 2.7 indicates. It is an X-ray image of the young (320 years) supernova remnant Cassiopeia A, whose structure is obviously clumpy. Barely visible is a point source of X-rays near the centre of the image, which may be the collapsed core of the star that became the supernova. Detailed abundance determinations require spatial and spectral resolution of the clumps. Combing all wavelengths to build a model of the nebula gives the abundance estimates we seek.

Figure 2.8 shows an image of Cas A taken with the Hubble Space Telescope. Different glowing filaments are found to have different chemical compositions: for example, some are rich in oxygen whereas others are rich in sulphur. The signature of the oxygen layer in the onion-skin model would be unburned oxygen

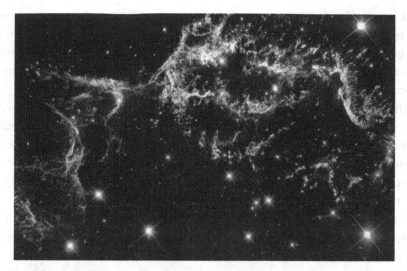

Figure 2.8 Cassiopeia A in optical light (NASA and the Hubble Heritage Team STScI/AURA; R. Fesen (Dartmouth) and J. Morse (University of Colorado)).

and burned oxygen (in this case sulphur, Hoyle 1954). Other regions can be identified that correspond to an oxygen shell exhibiting different stages of burning, from unburned to completely burned. It is a challenging puzzle to identify the original layers of the presupernova (Fesen 2001), and to simulate them correctly in a detailed model of the evolution up to and during explosion.

2.5   Neutronization, NSE and core collapse

Hoyle (1946) delineated the conditions of nuclear statistical equilibrium (NSE) in stars. This discussion became a key part of Burbidge, Burbidge, Fowler and Hoyle (1957) as well as of subsequent attempts to model the approach to NSE and its freeze-out (Truran, Cameron and Gilbert 1966, Truran Arnett and Cameron 1967).

The neutrino–antineutrino emission mechanism of Gamow and Schoenberg (1941) was considered by Hoyle in 1946, but it was shown that a more rapid energy loss was due to photodissociation of nuclei at high temperature. This implied that the trigger for hydrodynamic collapse, as opposed to quasi-hydrostatic contraction, would be photodissociation of iron-group nuclei into alpha particles (and some free nucleons), as in

$$^{56}\text{Fe} \leftrightarrow 13\alpha + 4\text{n}. \tag{2.1}$$

Later (Fowler and Hoyle 1964) this question was revisited, including a set of new processes, namely

$$e^- + e^+ \rightarrow \nu + \bar{\nu} \tag{2.2}$$

and its relatives, which were predicted on the basis of the conserved-vector-current (CVC) theory of Feynman and Gell-Mann (1958). The CVC theory was replaced by the electro-weak neutral-current theory of Weinberg, Glashow and Salam, but the rates calculated from the newer theory were roughly the same (Dicus 1972).

In rereading Hoyle (1946) it became clear that Hoyle had solved a problem that still puzzled astrophysicists 25 years later. As the stellar core contracts to increasingly higher density, the electrons become increasingly degenerate. It becomes energetically favourable to a radioactive neutron-rich nucleus and fewer electrons. The rising electron Fermi energy prevents the radioactive decay. Thus electron degeneracy modifies the nuclear statistical equilibrium. There develops a beta balance, in which thermal spreading allows electron capture and electron decay to move towards a steady state. Hoyle noted that this was not detailed balance (both neutrinos and antineutrinos escaped), but that the equilibrium was analogous to dissociation. Modern terminology would be that the chemical potentials of both neutrino and antineutrino would be zero because they escape freely; in radiative dissociation the photon plays the role of the neutrinos and antineutrinos, and has zero chemical potential. Using a Weizsäcker mass law from Bethe and Bacher (1936), Hoyle found a maximum degree of neutronization and the most abundant isotope for a given density. This solution was not improved upon until much later, e.g. by Lamb, Lattimer, Pethick and Ravenhall (1971), Baym, Bethe and Pethick (1971), and Bethe, Brown, Applegate and Lattimer (1971).

Hoyle (1946) used energy arguments to determine the trajectory of collapse in ρ–T diagrams, in reasonable agreement with subsequent work (see Arnett 1996 for detailed discussion and references). Although the weak-interaction neutral currents were then undiscovered, Hoyle (1946) chose a collapsing core near the Chandrasekhar mass. We now know that these processes cool the core to make its electrons degenerate, so that this choice is nearer the mark than the estimates in Fowler and Hoyle (1964) (which assume a polytropic and non-degenerate structure). This is ironic in that Fowler and Hoyle (1964) attempted to improve the earlier work by a better treatment of the neutrino cooling and a better mathematical representation of the core conditions, yet did not include the deviation of the core structure driven by the neutrino cooling.

2.6   Nucleosynthesis processes and freeze-out

Hoyle (1946) explicitly notes some difficulties. First, the reaction rates for intermediate nuclei were largely unknown. Second, a realistic model of a supernova must involve an ensemble of freeze-outs, hence requiring such rates.

Hoyle (1954) improved this situation, and developed a strikingly modern picture of the processes between hydrogen burning and NSE, all of which Hoyle (1946) termed the 'alpha process'.

Hoyle (1954) discusses the building of nuclei from C to Ni, or in present terminology, helium, carbon, neon, oxygen, and silicon burning. We shall reserve a separate section below for a discussion of the reaction rate for helium burning. The nucleosynthesis consisted first of the triple-alpha reaction to ^{12}C by

$$^4He + {}^4He \leftrightarrow {}^8Be \tag{2.3}$$

and

$$^4He + {}^8Be \rightarrow {}^{12}C. \tag{2.4}$$

As the ^{12}C abundance increases,

$$^4He + {}^{12}C \rightarrow {}^{16}O \tag{2.5}$$

begins to be effective, so that at low ^4He (near exhaustion of fuel), this reaction dominates, reducing the ^{12}C abundance. Depending upon the ratio of the rates, and hence the abundances of ^4He and ^{12}C, as well as temperature and density, the product can range from high ^{12}C to almost pure ^{16}O. Öpik (1951) found a similar behaviour using more approximate estimates of the reaction rates. The subsequent step, $^4He + {}^{12}C \rightarrow {}^{16}O$, is non-resonant at these temperatures, and therefore slow.

The next stage is carbon burning, with fusion of two ^{12}C with several exit channels, mostly

$$^{12}C + {}^{12}C \rightarrow {}^{20}Ne + \alpha, \tag{2.6}$$
$$\rightarrow {}^{23}Na + p. \tag{2.7}$$

The production of alpha particles and protons, which react more easily than carbon fuses, make the ultimate product of such burning less than obvious. The net result is a mixture of isotopes of Mg, Al, and Si, as well as ^{20}Ne and ^{23}Na.

The next stage is neon burning, which proceeds in a tricky way. First we have dissociation by

$$^{20}Ne \rightarrow {}^{16}O + \alpha, \tag{2.8}$$

with the newly produced α being captured according to

$$^{20}Ne + \alpha \rightarrow {}^{24}Mg. \tag{2.9}$$

The net result is

$$^{20}\text{Ne} + {}^{20}\text{Ne} \rightarrow {}^{16}\text{O} + {}^{24}\text{Mg}. \tag{2.10}$$

Because two ^{20}Ne are less bound than ^{16}O $+ {}^{24}$Mg, neon burning is slightly exoergic.

The next stage is oxygen burning, with fusion of two ^{16}O with several exit channels, mostly

$$^{16}\text{O} + {}^{16}\text{O} \rightarrow {}^{28}\text{Si} + \alpha, \tag{2.11}$$

$$\rightarrow {}^{31}\text{P} + \text{p}. \tag{2.12}$$

The production of alpha particles and protons, which react much more easily than oxygen fuses, make the ultimate product of such burning complicated and sensitive to the burning temperature. The net result is a range of isotopes from ^{28}Si to ^{40}Ca, using currently plausible temperatures. At lower temperatures the range is truncated to ^{28}Si and ^{32}S with various more minor isotopes. Because the cooling was underestimated, owing to lack of neutral-current neutrino emission processes, Hoyle (1955) obtained the smaller range, and struggled with Ar and Ca production. Explosive burning occurs at a higher temperature, and therefore eases this problem. The fusion of ^{12}C $+ {}^{16}$O was also considered; measurements of the cross sections now indicate that this is a minor channel.

While the sequence of carbon, neon and oxygen burning laid out by Hoyle (1955) is thought to be correct today, Burbidge, Burbidge, Fowler and Hoyle (1957) reverted back to the notion of an 'alpha process' to synthesize the nuclei resulting from these processes. They state 'no important processes occur among ^{12}C, ^{16}O, and ^{20}Ne until significantly higher temperatures, of the order of 10^9 degrees are attained.' With measurements of the carbon fusion rate, this was revised back to the Hoyle (1955) description.

After oxygen burning comes the approach to NSE (now called silicon burning – Truran, Cameron and Gilbert 1966 and Bodansky, Clayton and Fowler 1968). This was too difficult to tackle then, so the discussion focused on properties of α, γ chains from Si to Ti. NSE itself was re-examined, and for the first time theoretical and observed (meteoritic) abundance ratios were compared quantitatively (p. 144 in Hoyle 1955).

The Burbidge, Burbidge, Fowler and Hoyle (1957) paper is too famous to require detailed analysis here. It represented several improvements: use of the new Solar-System abundances of Suess and Urey (1956), more accurate nuclear physics, introduction of the s, r, and p processes, cosmochronology with U and Th, and more astronomical comparisons (such as the technetium observations of Merrill (1952) and stellar abundances by the Burbidges). The synthesis bears the marks of all the authors, and is an amazing contribution.

2.7   Thermonuclear runaway and supernovae type Ia

Following the observational distinction between Type I and Type II supernovae, Hoyle and Fowler (1960) divided theoretical explosions into two types
as well: the core collapse events discussed above, and a white-dwarf thermonuclear explosion. It now appears that the latter may correspond to an accreting
white dwarf, a growing stellar core in an intermediate-mass star (if it does not
save itself by 'superwind' mass loss), or merger of a binary pair of dwarfs. At
high electron degeneracy, thermonuclear fusion of C or O is greatly enhanced
by electron screening, that is the shielding cloud of electrons becomes so com-
pressed that it reduces the Coulomb repulsion of the nuclei. As the reactions
heat the plasma, pressure does not increase significantly because it is dominated by the electron-degeneracy pressure. Thus the temperature rises further,
the reactions speed up, and a thermal runaway ensues. In retrospect we can see
difficulties: how does the ignition happen, does a detonation or a deflagration
develop, and what exactly is the path the progenitor takes to get to instability? We still cannot answer these questions convincingly, despite the fact that
Type Ia supernovae are the foundation for the cosmological distance scale at
depth. We do not yet have the answers, but Hoyle and Fowler (1960) defined the
problem.

2.8   Red giants and levels in ^{12}c

The 1997 Craafoord Prize was awarded by the Swedish Academy of Sciences to Sir Fred Hoyle and E. E. Salpeter for 'pioneering contributions to the
study of nuclear processes in stars and stellar evolution'. It was an impressive
event, with families and grandchildren of the recipients. Much of the scientific discussion was about a seminal problem: what follows hydrogen burning?
Salpeter (1952) had calculated the rate for the triple-alpha reaction,

$$2\alpha \rightarrow {}^8\text{Be},$$ \hfill (2.13)

$$^8\text{Be} + \alpha \rightarrow {}^{12}\text{C},$$ \hfill (2.14)

assuming that the second stage was non-resonant. Hoyle and Schwarzschild
(1955) had explained the old problem of the red-giant branch seen in stellar
clusters as the slow growth of an inert helium core topped with a thin hydrogen-
burning shell and an extended envelope which was convective. The tip of the
giant branch seemed to be the point at which helium burning ignited in the
core. Consequently, the observed luminosity at the tip could be connected with
the density and temperature in the core. Unfortunately, Salpeter's rate was too
slow to agree with observations.

Fowler (1984, his Nobel Prize lecture) describes how this problem was solved. Hoyle arrived at Caltech with good reason to suspect that a resonance at the right energy might be the solution. Now a resonance at the appropriate energy had been suggested before, and was in the previous Lauritsen tables (Hornyak and Lauritsen 1948, Hornyak, Lauritsen, Morrison and Fowler 1950, and Ajzenberg and Lauritsen 1952). Sir Fred gave a seminar explaining the astrophysics problem. He was then told that the latest experiment using new and much improved techniques (Malm and Buechner 1951) did not see it, and the level was to be dropped from the tables. There are at least two types of theorist. At this point, most would simply adjust (or introduce) a parameter and happily produce results in agreement with the data. Fred was not this type; he is supposed to have replied, paraphrasing Eddington, that no experiment should be believed until proven by theory. Ward Whaling heard the seminar and quietly went to remeasure the reaction rate; the resonance appeared to be there after all (Dunbar, Pixley, Wenzel and Whaling 1953).

Arnett (1996, p. 223) stated that Hoyle predicted the level; actually he predicted that the newly expunged level would be real. I probably got the idea from Fowler's Nobel lecture; Fred never made any special claims on the subject to me.

2.9   Conclusions

Fred said that nucleosynthesis preceded his Steady-State cosmology, and the literature bears him out on this. He felt that if making the elements required no special conditions other than those found in stars, then a new set of cosmological possibilities arose.

Many of Hoyle's contributions were not attributed to him, but taken as already given and not needing justification or reference, probably because he produced them so early. In rereading his work, I found myself guilty of this.

Fred Hoyle named the Big Bang, and believed it to be wrong. However, he took part in the first detailed examination of the nucleosynthesis predictions of this cosmology (Wagoner, Fowler and Hoyle 1967), and there can be no doubt that he was a dominant participant in the debate that drove cosmology to the point of becoming a quantitative science.

Two Nobel Prize citations concern nucleosynthesis in stars: Bethe (1967) and Chandrasekhar and Fowler (1983). Second-guessing the Swedish Academy of Science is inappropriate. The Nobel is their prize, and the awardees are almost all unquestionably deserving (that is a laudable record for a merely human institution). However, the topic I have addressed here demands some assessment of Hoyle's contribution to the field, and it is clear that his contributions are at that high standard.

Shortly after learning of the Nobel award, at breakfast at a nucleosynthesis conference at Yerkes Observatory, Fowler said, 'The only thing that worries me is how will Fred take all this?' It seems that Fowler would have felt less awkward if Fred had had a part of the Prize too.

Acknowledgments

Grateful acknowledgments are due to Virginia Trimble and Don Clayton for helpful discussions. I am indebted to Dr Louis Brown who provided a copy of his interesting correspondence with W. A. Fowler concerning the second excited state in ^{12}C and the stability of ^{8}B. Support by DOE under grant number DE-FG03-98DP00214/A001, and a subcontract from ASCI Flash Center at the University of Chicago, is gratefully acknowledged.

References

AJZENBERG, F. & LAURITSEN, T. 1952 *Rev. Mod. Phys.*, **24**, 321

ALPHER, R. A. & HERMAN, R. C. 1953 *Ann. Rev. Nucl. Sci.*, **2**, 1

ARNETT, D. 1974 *ApJ*, **195**, 727

—— 1996 *Supernovae and Nucleosynthesis* (Princeton: Princeton University Press)

BAYM, G., BETHE, H. A. & PETHICK, C. J. 1971 *Nucl. Phys. A*, **175**, 225

BETHE, H. A. & BACHER, R. F. 1936 *Rev. Mod. Phys.*, **8**, 82 and 165

BETHE, H. A., BROWN, G. E., APPLEGATE, J. & LATTIMER, J. M. 1971 *Nucl. Phys. A*, **324**, 487

BLOOM, J. S. *et al.*, 2002 *ApJL*, **572**, 45

BODANSKY, D., CLAYTON, D. D. & FOWLER, W. A. 1968 *ApJS*, **16**, 299

BONDI, H. & HOYLE, F. 1944 *MNRAS*, **104**, 27

BURBIDGE, E. M., BURBIDGE, G. R., FOWLER, W. A. & HOYLE, F. 1957 *Rev. Mod. Phys.*, **29**, 547

CHANDRASEKHAR, S. & HENRICH, L. R. 1942 *ApJ*, **95**, 288

DICUS, D. 1972 *Phys. Rev. D*, **6**, 941

DUNBAR, D. N. F., PIXLEY, R. E., WENZEL, W. A. & WHALING, W. 1953 *Phys. Rev.*, **92**, 64

FESEN, R. A. 2001 *ApJS*, **133**, 161

FEYNMAN, R. P. & GELL-MANN, M. 1958 *Phys. Rev.*, **109**, 193

FOWLER, W. A. 1984 *Rev. Mod. Phys.*, **56**, 149

FOWLER, W. A. & HOYLE, F. 1964 *ApJS*, **9**, 201

FRYXELL, B. A., MÜLLER, E. & ARNETT, D. 1991 *ApJ*, **367**, 619

GAMOW, G. & SCHOENBERG, M. 1941 *Phys. Rev.*, **59**, 539

GARNAVICH, P. M., *et al.* 2003 *ApJ*, **582**, 924

HORNYAK, W. F. & LAURITSEN, T. 1948 *Rev. Mod. Phys.*, **20**, 191

HORNYAK, W. F., LAURITSEN, T., MORRISON, P. & FOWLER, W. A. 1950 *Rev. Mod. Phys.*, **22**, 291

HOYLE, F. 1945 *MNRAS*, **105**, 287 and 302
  1946 *MNRAS*, **106**, 366
  1948 *MNRAS*, **108**, 372
  1954 *ApJS*, **1**, 121
  1955 *Frontiers of Astronomy* (New York: Harper & Brothers)
HOYLE, F. & FOWLER, W. 1960 *ApJ*, **132**, 565
HOYLE, F. & LYTTLETON, R.A. 1941 *MNRAS*, **101**, 227
HOYLE, F. & SCHWARZSCHILD, M. 1955 *ApJS*, **2**, 1
LAMB, D.Q., LATTIMER, J.M., PETHICK, C.J. & RAVENHALL, D.G. 1971
  Phys. Rev. Lett., **41**, 1623
MALM, R. & BUECHNER, W.W. 1951 *Phys. Rev.*, **81**, 519
MERRILL, P.W. 1952 *Science*, **15**, 484
ÖPIK, E.J. 1951 *Proc. Roy. Irish Acad.*, **A54**, 49
PATAT, F. *et al.* 2001 *ApJ*, **555**, 900
PIRO, L. 2001 *GCN Circ.*, 1147
PRICE, P.A. *et al.* 2002 *ApJL*, **572**, 51
SALPETER, E.E. 1952 *ApJ*, **115**, 326
SUESS, H.E. & UREY, H.C. 1956 *Rev. Mod. Phys.*, **28**, 53
TRURAN, J.W., CAMERON, A.G.W. & GILBERT, A. 1966 *Can. J. Phys.*, **44**, 563
TRURAN, J.W., ARNETT, D. & CAMERON, A.G.W. 1967 *Can. J. Phys.*, **45**, 2315
WAGONER, R.V., FOWLER, W.A. & HOYLE, F. 1967 *ApJ*, **148**, 3
WOOSLEY, S.E. & WEAVER, T.A. 1981 *ApJ*, **243**, 561

3

Fred Hoyle: contributions to the theory of galaxy formation

GEORGE EFSTATHIOU

Institute of Astronomy, University of Cambridge

I review two fundamental contributions that Fred Hoyle made to the theory of galaxy formation. Hoyle was the first to propose that protogalaxies acquired their angular momentum via tidal torques from neighbouring perturbations during a period of gravitational instability. To my knowledge, he was also the first to suggest that the masses of galaxies could be explained by the requirement that primordial gas clouds cool radiatively on a suitable timescale. Tidal torques and cooling arguments play a central role in the modern theory of galaxy formation. It is a measure of Hoyle's breadth and inventiveness that he recognized the importance of these processes at such an early stage in the history of the subject.

3.1   Introduction

I begin by quoting from an obituary of Sir Fred Hoyle written by Leon Mestel (2001):

> Fred Hoyle was the astrophysicist *par excellence*, and much else. He wrote technical papers on an astonishingly wide range of astronomical topics, his most important work permanently widening our vistas and influencing strongly the direction of future research.

The 'much else' referred to by Leon in this quote includes Hoyle's contributions to the popularization of science, prolific science writing, his creation and

The Scientific Legacy of Fred Hoyle, ed. D. Gough.
Published by Cambridge University Press. © Cambridge University Press 2004.

directorship of the Institute of Astronomy at Cambridge and his often visionary work for the UK Science Research Council.

In this volume our focus is on Hoyle's astronomical research. Everyone would agree that Hoyle's outstanding scientific achievements were in developing the theory of stellar evolution and in understanding the origin of the chemical elements. As appropriate, these topics have taken the central stage. Hoyle's contributions to galaxy formation, the subject of this chapter, are less well known. I have chosen this topic to illustrate the 'astonishingly wide range' of Hoyle's research referred to by Leon Mestel. I shall argue that Hoyle was the first to understand the masses of galaxies and why galaxies rotate. These are important results in their own right and now form an essential part of the modern theory of galaxy formation. Many astronomers would be proud to list these two results (which are not even mentioned in any of Hoyle's obituaries) as their towering achievements!

Section 3.2 discusses the origin of galactic angular momentum, and is based on Hoyle's (1949) contribution to the quaintly named volume *Problems of Cosmical Aerodynamics* edited by Burgers and van de Hulst. This volume summarizes the proceedings of the equally quaintly named *Symposium on the Motion of Gaseous Masses of Cosmical Dimensions* held in Paris in 1949. Section 3.3 discusses the origin of galactic masses. This is based on Hoyle's well-known paper 'On the fragmentation of gas clouds into galaxies and stars' (Hoyle 1953). Curiously, most of the citations to this paper refer to the sections on star formation (hierarchical, opacity limited, fragmentation), yet this paper provides a clear account of the cooling–timescale explanation of galaxy masses, predating the work of Binney (1977), Rees and Ostriker (1977) and Silk (1977) by more than 20 years.

3.2   Origin of galactic rotation

Consider the situation shown in Figure 3.1. Here we have an ellipsoid separated by distance r from an object of mass M. Let us assume that the

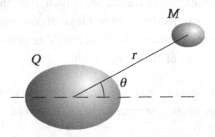

Figure 3.1 An ellipsoid with quadrupole moment Q separated by distance r from a mass M.

ellipsoid is a homogeneous oblate spheroid of density ρ, mass M_G, semi-major axis a and semi-minor axis b. The magnitude of the torque on the spheroid is $(r \gg a)$

$$\Gamma = \frac{3GM}{2r^3}(\sin 2\theta)\rho \int (x^2 - z^2)\mathrm{d}^3\mathbf{x}$$

$$= \frac{3GMQ}{4r^3}\sin 2\theta, \qquad Q = \frac{2}{5}M_G(a^2 - b^2), \tag{3.1}$$

where Q is the quadrupole moment of the spheroid.

How does Hoyle apply equation (3.1)? First, we need to estimate the timescale over which the torque acts. For this, Hoyle argues that the torque will act on a timescale comparable to the collapse timescale of a protogalaxy $(\sim(3/4\pi G\rho)^{1/2})$, where for ρ Hoyle adopts the value 10^{-27} g/cm^3. This was Hoyle's estimate of the present mean matter density of the Universe and is high, presumably because estimates of the Hubble constant were so high at the time (see for example the discussion of the age discrepancy of evolving world models in Bondi and Gold 1948). We can re-interpret Hoyle's estimate as the mean density of the Universe at the redshift at which the protogalaxy begins to form, $(1 + z_f) \approx 4(\Omega_m h^2)^{-1/3}$, which nowadays we would regard as a reasonable number. Hoyle then assumes that the protogalaxy collapses until it is centrifugally supported, in which case the final angular velocity will be

$$\Omega_f \approx \left(\frac{r^3}{3MGx}\right)^3 \left(\frac{4\pi G\rho}{3}\right)^{7/2}, \qquad x = \frac{5Q}{4M_G a^2}\sin 2\theta. \tag{3.2}$$

To estimate the term (r^3/M) in the first of equations (3.2) Hoyle uses a clever argument. I quote directly from his paper:

> We now reach the important step of interpreting the external gravitational field that produces the couple acting on the condensation. Instead of regarding this field as arising from a neighbouring galaxy, we notice that there are large scale irregularities in the distribution of the internebular material. The existence of such irregularities probably exist also among the general field nebula, as is evidenced by the occurrence of peculiar velocities among these galaxies. Since the peculiar velocities average about 200 km s^{-1} in the neighbourhood of our own galaxy, this indicates $(GM/r)^{1/2} \approx 2 \times 10^7$ cm s^{-1} in this neighbourhood. In addition the observed over-all radii of the great nebular clusters are of order 10^6 parsecs, which suggests a value of r of order 3×10^{24} cm.

Given how little was known about galaxy clustering at the time, this is a brilliant way of constraining the matter distribution, and leads to Hoyle's final estimate of

$$\Omega_f \approx \frac{10^{-16}}{x^3}\ s^{-1}, \tag{3.3}$$

which he argues is consistent with the characteristic angular velocities of the Milky Way and Andromeda ($\sim 10^{-15}\ s^{-1}$) if the dimensionless number x is of order 1/3 (a reasonable value).

What did the conference participants make of this argument? There is a wonderful discussion reported at the end of Hoyle's contribution, which I have abstracted:

> *Heisenberg*: I had some difficulty understanding Dr Hoyle's
> argument...You start with an irregular cloud and then you ask why
> this irregular cloud gets an angular momentum. Now, I don't think
> that any theory can give us an irregular cloud from the beginning
> without giving to it angular momentum at the same time...If you
> have an irregular cloud, then it must have been produced by some
> kind of astronomical turbulence.[1]
>
> *Hoyle*: I take it that Professor Heisenberg assumes turbulence to be
> present in the intergalactic medium arising from some unknown
> source. For my part I would regard the gravitational forces as the
> basic phenomena. The energy released by the gravitational
> contraction might possibly furnish turbulent motions and might
> lead to the appearance of eddies. I feel very doubtful, however, about
> the assumption that the intergalactic medium should already have
> turbulence from itself.
>
> *Heisenberg*: May I express my view in the following way. A cloud means
> turbulence and thus you should not start with a sphere without
> turbulence, since anything which is a cloud is turbulent by itself.
>
> *Batchelor*: [Who understood Kelvin's circulation theorem] Why?
>
> *Heisenberg*: How can an irregular thing like a cloud have originated
> other than as a consequence of turbulent motion?
>
> *Hoyle*: A cloud can form in a more-or-less uniform medium through
> gravitational instability.

I can find nothing of significance on the tidal-torque theory in the literature (apart from Sciama's 1955 application to galaxy formation in the Steady-State

---

[1] By 1949 Heisenberg's research interests had shifted to the study of turbulence. It is rumoured that Heisenberg said that he was looking forward to discussing quantum mechanics with other physicists in Heaven after he was dead. However, he thought that he would need a personal audience with God to understand turbulence.

theory) until Peebles' important paper in 1969. In fact, the short abstract of Peebles' paper almost echoes Hoyle's last remark:

> It is shown that the angular momentum of rotation of the Galaxy agrees in magnitude with the prediction of the gravitational instability picture for the rotation of the galaxies.

In this paper, Peebles analysed the tidal-torque theory using linear perturbation theory, and showed that it could account, roughly, for the angular momentum of the Milky Way. As part of my thesis work, I used N-body simulations to estimate the efficiency of the tidal-torque mechanism (Efstathiou and Jones 1979). These simulations suggested that the dimensionless spin parameter λ (roughly the ratio of rotational to kinetic energy) for a protogalaxy would have a low value of

$$\lambda = J|E|^{1/2}G^{-1}M^{-5/2} \approx 0.05, \tag{3.4}$$

a result that was later borne out by much larger numerical simulations (Barnes and Efstathiou 1987, Zurek, Quinn and Salmon 1988). However, such a low value of λ leads to problems with the theory as envisaged by Hoyle and Peebles, for if a self-gravitating gas cloud collapses from an initial radius R_i to form a centrifugally supported exponential disc of scale-length α, it must collapse by a huge factor of

$$\alpha R_i \approx 0.7/\lambda^2 \approx 300, \tag{3.5}$$

if $\lambda \approx 0.05$. The collapse time for a typical spiral disc would therefore be longer than the age of the Universe; hence the theory is untenable. A solution to this problem was proposed by Efstathiou and Jones (1980) and worked out in detail by Fall and Efstathiou (1980). If spiral discs form from the collapse of baryonic material within a dissipationless dark halo (as envisaged in the two-component theory of White and Rees 1978), and provided that the gas conserves its angular momentum and is only marginally self-gravitating when it reaches centrifugal equilibrium, then it only needs to collapse by a factor of

$$\alpha R_i \approx \sqrt{2}/\lambda \approx 30, \tag{3.6}$$

if $\lambda \approx 0.05$.

Does this lead to a viable theory of galaxy formation? Many cosmologists would probably say yes, but there are some thorny issues that have not yet been resolved. The first gas-dynamics simulations of galaxy formation showed that much of the gas collapsed into dense sub-units at early times (as expected from the over-cooling problem identified by White and Rees 1978 and discussed in more detail in the next section). These sub-units lose their orbital angular momentum by dynamical friction as they merge to form larger sub-units. The

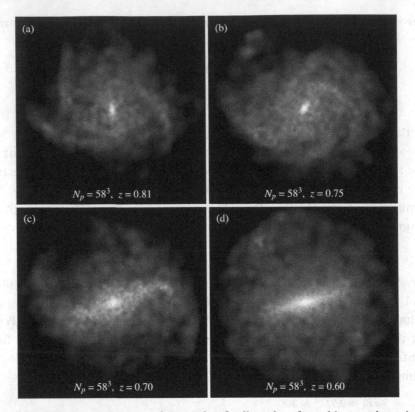

Figure 3.2 Face-on images at four epochs of a disc galaxy formed in a gas-dynamical numerical simulation. Each picture shows a square patch of size 20 kpc × 20 kpc.

end result is a blob of gas or stars with specific angular momentum one or two orders of magnitude lower than observed in real disc galaxies[2] (see e.g. Navarro and White 1994, Navarro and Steinmetz 1997). One possible solution to this problem is to invoke feedback from supernovae to prevent the gas from collapsing at early times. (There are differing views: for example, Governato *et al.* (2004) argue that the catastrophic angular momentum loss found in early simulations is caused by their limited numerical resolution.) Recently, numerical simulations that include simplified models of stellar feedback have shown that it is possible to make disc systems with specific angular momenta and scale-lengths comparable to those of L^* galaxies (Weil, Eke and Efstathiou 1998, Sommer-Larsen, Götz and Portinari 2002, Abadi *et al.* 2003). An example is shown in Figure 3.2 (Wright, Efstathiou and Eke 2003). This simulation begins with scale-invariant adiabatic initial conditions in a Λ-dominated cold dark matter (CDM) cosmology

---

[2] This possibility was first pointed out to me by Peter Goldreich in 1981.

($\Omega_m = 0.3$, $\Omega_\Lambda = 0.7$). The gas is (artificially) prevented from cooling before a redshift of unity, but once cooling sets in, the gas collapses to form a disc within the dark halo preserving a large fraction of its angular momentum. As one can see from the figure, the disc displays (transient) spiral arms at early times and eventually forms a strong bar. This richness of structure is a direct result of the angular momentum acquired by tidal torques during the early stages of gravitational instability, just as Hoyle predicted more than 50 years ago.

3.3   Explaining galaxy masses

Shortly after the discovery of the expansion of the Universe, it was re-alized that galaxies had a well-defined upper luminosity (Hubble and Humason 1931). The distribution of galaxy luminosities (the galaxy luminosity func-tion) was subsequently studied by many authors (see e.g. Binggeli, Sandage and Tammann 1988) and has recently been estimated with extremely high precision using the 2dF and SDSS galaxy redshift surveys (Blanton *et al.* 2001, Norberg *et al.* 2002). These studies show that the galaxy luminosity function is well described by a Schechter (1976) function and that 90% of the mean stel-lar luminosity density is contained in galaxies spanning the luminosity range $0.02 \lesssim (L/L^*) \lesssim 2.5$, where $L^* \approx 2.6 \times 10^{10} L_\odot$ in the b-band. Most of the stellar luminosity in the Universe is therefore confined to galaxies covering a range of about a hundred or so in luminosity. Furthermore, there is compelling evidence for a critical stellar mass of about $3 \times 10^{10} M_\odot$ (Kauffmann *et al.* 2002); galaxies of higher mass have roughly similar central surface brightnesses, independent of mass or luminosity, while galaxies with lower stellar masses have lower surface brightnesses with $\langle \mu \rangle \propto M^{0.54}$.

Clearly, some physical process is required to explain the restricted range of galaxy luminosities and masses. In Hoyle's own words, one of the aims of his 1953 paper was to explain 'Why are the masses of galaxies mainly confined in the range $3 \times 10^9 M_\odot$ to $3 \times 10^{11} M_\odot$, with possibly a tendency to fall into two groups at the end of this range?'. Hoyle's explanation was based on simple gas dynamics – the requirement that a proto-cloud of hydrogen, at a few times the mean cosmic density, be able to cool radiatively on a timescale shorter than the timescale for gravitational collapse.

Hoyle's argument can be illustrated in a way more easily accessible to mod-ern readers using the famous cooling diagram, reproduced in Figure 3.3, from Rees and Ostriker (1977). A gas cloud with a particle density of $n \sim 10^{-3} \, \text{cm}^{-3}$ ($\rho \sim 10^{-27} \, \text{g/cm}^3$, as considered by Hoyle) intersects the solid curve delineating region C in the diagram at two points, as indicated by the heavy vertical line in the figure. Lines of constant Jeans mass (with slope 1/3 in this diagram) are

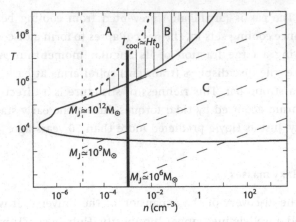

Figure 3.3 Within region C, bounded by the solid curve, a gas cloud of uniform density n, temperature T, and primordial composition can cool within a free-fall time. Clouds within region B can collapse quasi-statically within a Hubble time. Clouds in region A cannot cool within a Hubble time (From Rees and Ostriker, 1977).

shown by the dashed lines. Evidently a density of $n \approx 10^{-3}\,\mathrm{cm^{-3}}$ defines a lower Jeans mass of $\sim 5 \times 10^8\,M_\odot$ and an upper Jeans mass of $\sim 10^{12}\,M_\odot$. The interpretation of this result is as follows: region C delineates the region of the T–n plane in which gas clouds can cool radiatively within a free-fall time. Thus, clouds with mean density $n \sim 10^{-3}\,\mathrm{cm^{-3}}$ heated to their virial temperatures by gravitational collapse can cool and achieve the high overdensities ($\gg 10^5$) of normal galaxies only if their masses lie within the range $5 \times 10^8\,M_\odot \lesssim M \lesssim 10^{12}\,M_\odot$. These numbers differ from those in Hoyle's paper because Hoyle neglected cooling by line emission, and he did not phrase the argument in terms of the virial temperatures of protoclouds in quite the way discussed here. The key point, however, that cooling times delineate upper and lower bounds for galaxy masses, is contained in Hoyle's paper.

Does this cooling argument really explain galaxy masses? Since we no longer believe in the Steady-State theory, we can ask what happens to clouds with densities much higher than $n \sim 10^{-3}\,\mathrm{cm^{-3}}$. If the virial temperature of such a cloud exceeds $10^4\,\mathrm{K}$, then it can cool efficiently on a timescale much shorter than a free-fall timescale. The lower mass limit of Hoyle's argument is 'soft' because it is sensitive to the redshift at which protogalaxies collapse.

In the cold dark matter (CDM) model, this leads to a cooling catastrophe, as mentioned in the previous section. In the absence of additional heating sources, the baryonic material would be expected to collapse efficiently in small non-linear systems with virial temperatures $\gtrsim 10^4\,\mathrm{K}$ at high redshifts (White and Rees 1978). According to the Press–Schechter (1974) theory, in a hierarchical

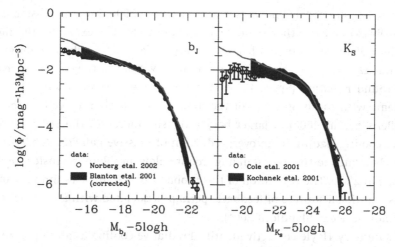

Figure 3.4 Semi-analytic models of the galaxy luminosity function from Cole
et al. (2000) (solid lines) compared with observations of the b-band
(left-hand panel) and K-band (right-hand panel) luminosity functions
(Figure from Baugh *et al.* 2003).

model with a power-law spectrum of fluctuations $P(k) \propto k^n$, the mass function
of dark haloes at low masses varies as

$$\frac{dN(m)}{dm} \propto m^{-(9-n)/6}. \tag{3.7}$$

For a CDM model, $n \approx -2$ on the scales relevant for galaxy formation, and so
the 'cooling catastrophe' would lead to a much steeper mass spectrum than that
inferred from the faint-end slope of the galaxy luminosity function ($dN(L)/dL \propto$
$L^{-1.2}$). A photoionizing background can prevent the gas from cooling in
haloes with circular speeds less than about 30 km s^{-1} (Efstathiou 1992, Benson
et al. 2002) and this can help prevent the formation of dwarf galaxies. However, most authors are agreed that substantial feedback from supernovae-driven
winds is needed to reproduce the shape of the galaxy luminosity function in
CDM models (see e.g. White and Frenk 1991, Cole *et al.* 2000). As with the angular-
momentum problem discussed in the previous section, it seems that stellar feedback is required to explain the observed properties of galaxies in a CDM universe.

This is illustrated in Figure 3.4 from Baugh *et al.* (2002). The left-hand panel
shows estimates of the b-band luminosity function from Blanton *et al.* (2001) and
Norberg *et al.* (2002), while the right-hand panel shows estimates of the K-band
luminosity function from Cole *et al.* (2001). The solid lines in the figure show
the predictions of the semi-analytic models of Cole *et al.* (2000), which include
substantial stellar feedback. These models provide an acceptable match to the observations, but the model predictions are sensitive to the details of the feedback

model which are poorly known (see e.g. Efstathiou 2000, for a discussion of stellar feedback) and to other cosmological parameters. For example, the models of Cole *et al.* (2000) assume a baryon density of $\Omega_b = 0.02$, which is lower than the value $\Omega_b \approx 0.045$ inferred from recent observations of the cosmic microwave background radiation (Spergel *et al.* 2003). A higher baryon density introduces problems with the bright end of the luminosity function as well as the faint end (Benson *et al.* 2003). A larger baryon density increases the efficiency of radiative cooling leading to an overproduction of massive galaxies. (This problem is closely related to the long-standing 'cooling flow' problem in cluster cores, see e.g. Fabian 2002). The resolution of this problem is still unclear and probably requires additional sources of feedback, perhaps associated with a massive central black hole (see e.g. Blandford 2001).

In summary, Hoyle correctly identified radiative cooling as an important process in determining the baryonic masses of galaxies. However, 50 years after Hoyle's paper, we still do not understand many of the physical processes that produce galaxies from an almost featureless spectrum of density fluctuations. Cooling clearly plays a role, but so does energy injection from stars and possibly AGN. The physics of these feedback mechanisms is poorly understood, and yet is crucial for understanding galaxy formation. As Figure 3.5 shows, this physics is vital, for without it we cannot relate the dark-matter distribution to the visible-mass distribution at the present day and at higher redshifts.

3.4   Conclusions

Hoyle's explanation of the rotation of galaxies and his recognition of the importance of radiative cooling in fixing the baryonic masses of galaxies are not usually listed amongst his most important contributions to astronomy. Yet these processes are central to modern theories of galaxy formation. It is a measure of Hoyle's breadth and inventiveness that he recognized the importance of these processes at such an early stage in the subject. Nevertheless, more than 50 years after Hoyle's papers there are still major gaps in the theory of galaxy formation. This is because complex physical effects, in particular energy injection from massive stars, are important in determining the baryonic content and angular momentum properties of galaxies.

I will end with a quotation from Fred Hoyle's book *Galaxies, Nuclei and Quasars*, published in 1966:

> It is not too much to say that the understanding of why there are these different kinds of galaxy, of how galaxies originate, constitutes the biggest problem in present-day astronomy. The properties of the

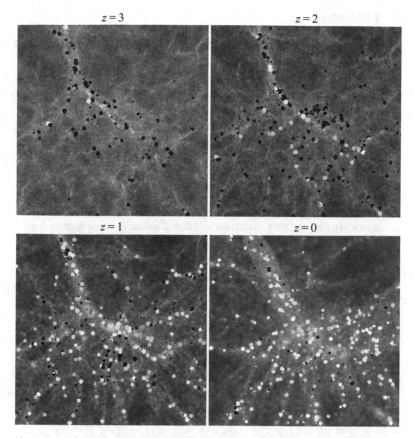

Figure 3.5 Slices through a Λ-dominated dark-matter simulation at various
redshifts. Simple prescriptions from semi-analytic models have been
applied to compute the properties of visible galaxies (masses, stellar ages,
etc.), which are represented by the dark (young galaxies) and light (old
galaxies) spots (See Kauffmann *et al.* 1999).

individual stars that make up the galaxies form the classical study of
astrophysics, while the phenomena of galaxy formation touches on
cosmology. In fact, the study of galaxies forms a bridge between
conventional astronomy and astrophysics on the one hand, and
cosmology on the other.

This remains as true today as it was nearly 40 years ago.

Acknowledgments

I thank Lisa Wright and Vince Eke for allowing me to reproduce Figure 3.2, and Carlton Baugh for providing a copy of Figure 3.4.

References

ABADI, M. G., NAVARRO, J. F., STEINMETZ, M. & EKE V. R. 2003 *ApJ*, **597**, 21

BARNES, J. & EFSTATHIOU, G. 1987 *ApJ*, **319**, 575

BAUGH, C. M., BENSON, A. J., COLE, S., FRENK, C. S. & LACEY, C. 2003
in *The Mass of Galaxies at Low and High Redshift*, eds. R. Bender & A. Renzini,
(Berlin/Heidelberg: Springer-Verlag) p. 91

BENSON, A. J., LACEY, C. G., BAUGH, C. M., COLE, S. & FRENK, C. S.
2002 *MNRAS*, **330**, 156

BENSON, A. J., BOWER, R. G., FRENK, C. S., LACEY, C. G., BAUGH, C. M.
& COLE, S. 2003 *ApJ*, **599**, 38

BINGGELI, B., SANDAGE, A. & TAMMANN, G. A. 1988 *ARA&A*, **26**, 509

BINNEY, J. 1977 *ApJ*, **215**, 483

BLANDFORD, R. 2001 in *Galaxies and their Constituents at the Highest Angular
Resolutions* (Proceedings of IAU Symposium 205) eds. R. T. Schilizzi *et al.*, p. 10

BLANTON, M. R. *et al.* 2001 *AJ*, **121**, 2358

BONDI, H. & GOLD, T. 1948 *MNRAS*, **108**, 252

COLE, S., LACEY, C. G., BAUGH, C. M. & FRENK, C. S. 2000 *MNRAS*, **319**, 168

COLE, S. *et al.* 2001 *MNRAS*, **326**, 255

EFSTATHIOU, G. 1992 *MNRAS*, **256**, 43p
2000 *MNRAS*, **317**, 697

EFSTATHIOU, G. & JONES, B. J. T. 1979 *MNRAS*, **186**, 133
1980 *Comments Ap. Space Sci.*, **8**, 169

FABIAN, A. C. 2002 in *Galaxy Evolution: Theory and Observations*, eds. V. Avila-Reese,
C. Firmani, C. Frenk & C. Allen, Revista Mexicana Astronomia Astrofisica (Ser.
Conf.), **17**, 303

FALL, S. M. & EFSTATHIOU, G. 1980 *MNRAS*, **193**, 189

GOVERNATO, F. *et al.* 2004 *ApJ*, **607**, 688

HOYLE, F. 1949 in *Problems of Cosmical Aerodynamics*, Proceedings of the Symposium
on the Motion of Gaseous Masses of Cosmical Dimensions, (Ohio: Central Air
Documents Office) 195
1953 *ApJ*, **118**, 513
1966 *Galaxies, Nuclei and Quasars* (London: Heinemann)

HUBBLE, E. & HUMASON, M. L. 1931 *ApJ*, **74**, 43

KAUFFMANN, G., COLBERG, J. M., ANTONALDO, D. & WHITE, S. D. M.
1999 *MNRAS*, **303**, 188

KAUFFMANN, G. *et al.* 2003 *MNRAS*, **341**, 54

MESTEL, L. 2001 *Astron. Geophys.*, **42**, 5.23

NAVARRO, J. F. & STEINMETZ, M. 1997 *ApJ*, **478**, 13

NAVARRO, J. F. & WHITE, S. D. M. 1994 *MNRAS*, **267**, 401

NORBERG, P. *et al.* 2002 *MNRAS*, **336**, 907

PEEBLES, P. J. E. 1969 *ApJ*, **155**, 393

PRESS, W. H. & SCHECHTER, P. 1974 *ApJ*, **187**, 425

REES, M. J. & OSTRIKER, J. P. 1977 *MNRAS*, **179**, 541

SCHECHTER, P. 1976 *ApJ*, **203**, 297

Sciama, D. W. 1955 *MNRAS*, **115**, 3

Silk, J. 1977 *ApJ*, **211**, 638

Sommer-Larsen, J., Götz, M. & Portinari, L. 2002 *Ap&SS*, **281**, 519

Spergel, D. N. *et al.* 2003 *ApJS*, **148**, 175

Weil, M. L., Eke, V. R. & Efstathiou, G. 1998 *MNRAS*, **300**, 773

White, S. D. M. & Frenk, C. S. 1991 *ApJ*, **379**, 52

White, S. D. M. & Rees, M. J. 1978 *MNRAS*, **183**, 341

Wright, L., Efstathiou, G. & Eke V. R. 2003 unpublished

Zurek, W. H., Quinn, P. J. & Salmon, J. K. 1988 *ApJ*, **330**, 519

4

Highlights of Fred Hoyle's work on interstellar matter and star formation

PHILIP M. SOLOMON

Department of Physics and Astronomy,
Stony Brook, State University of New York

4.1 Introduction

I first met Fred Hoyle in the summer of 1967 when I was a visitor to the brand new Institute of Theoretical Astronomy (IOTA), the precursor to the Institute of Astronomy. Fred had invited me to spend the summer in Cambridge at the suggestion of Peter Strittmatter. I remember the excitement near the end of the summer when the wonderful modern Institute building opened, and I think Fred would be very pleased with the expansion of the Hoyle Building just completed, and the tasteful way in which it was carried out. Although I was not working on either nuclear astrophysics or cosmology, his major research topics at the time, Fred took an interest in my work, and we developed a friendship. His interests were very broad, and he asked me to organize a meeting at the Institute in 1969 on the then new topic of infrared astronomy.

Over the years we discussed a huge variety of topics, ranging over infrared astronomy, the microwave background, cosmology, stellar mass loss, interstellar matter, comets, Stonehenge, the navigation system of birds, cricket and politics, particularly Watergate. When he came for a visit you never knew what the topic would be; Fred was not only stimulating and brilliant he was also fun to be with.

Fred Hoyle's work on interstellar matter and star formation can be divided into four periods spanning the years from 1940 until the beginning of the twenty-first century. The second period, in the early 1950s, includes his famous

The Scientific Legacy of Fred Hoyle, ed. D. Gough.
Published by Cambridge University Press. © Cambridge University Press 2004.

work on star formation (and galaxy formation) through fragmentation of gas clouds. The third period, extending from the 1960s to the 1970s, addresses the origin and composition of interstellar grains. The early work on grains (almost all with N. C. Wickramasinghe) included the first suggestion that grains formed in stellar atmospheres and were composed primarily of graphite and organic compounds including aromatic hydrocarbons rather than ice. Beginning about 1978 in a long series of papers and books (also with N.C.W.) Fred worked on the identification of interstellar grains with biological material and the influence of interstellar grains on the origin of life on Earth (panspermia).

Even though I knew Fred Hoyle pretty well, there were still a few surprises in the literature which I found in the process of preparing for the meeting. Long before his 1953 paper on the collapse and fragmentation of interstellar clouds, Fred Hoyle wrote a stunning article on the physics of interstellar gas in 1940 emphasizing the importance of molecular hydrogen. This was one of his first astronomical publications and his first on the physics of interstellar matter.

In addition to contributing to the literature on interstellar matter and star formation, from the early 1970s Fred was also actively interested in the new field of millimetre astronomy and the study of molecular clouds. In fact, in 1972, while still at the Institute, in a little-known effort he almost succeeded in obtaining funding for a UK millimetre-wave telescope to be located in Australia for the study of the Galactic Centre region and the southern Milky Way. Much later, after substantial modification, I believe, this led to the funding of the JCMT submillimetre telescope in Hawaii.

Here I shall discuss highlights from Fred's published work on interstellar matter from 1940 until about 1977 (elsewhere in this volume, Wickramasinghe gives a more comprehensive summary of the work on interstellar grains including work during the last 20 years).

4.2   The 1940 paper: 'on the physical aspects of accretion by stars'

This paper (Hoyle and Lyttleton 1940) contains the most complete treatment of the heating and cooling of interstellar gas published at that time. The context is the physical state of a pure hydrogen interstellar cloud accreting onto a massive star. In part, this was a reply to a criticism by Atkinson (1940) of their first (1939) accretion paper. For a pure hydrogen cloud the temperature had been previously estimated to be $10\,000\,K$, too high for substantial accretion. The problem was posed in terms of cooling mechanisms, and the solution, cooling by molecular hydrogen, H_2, was almost 30 years ahead of its time. The paper considers cooling by H_2, the photoionization and dissociation of H_2 and the formation of H_2. It is fascinating to see how Fred Hoyle formulated the problem more than 60 years ago even in the absence of any laboratory (or good theoretical) cross section or radiative lifetime.

4.2.1   Cooling by H_2

The estimate of 10,000 K depends on a calculation by Eddington based on the assumption that, apart from the radiation emitted in the capture of electrons by positive ions, the interstellar material has effectively no opportunity for further emission. It is therefore necessary to investigate whether the cosmical cloud is capable of emitting radiation by any process other than the inverse photo-electric effect. It will be shown that there is another process by which it can get rid of practically all the internal energy supplied by selective absorption...

...it can be seen that the main radiative processes are as follows:

(1)   Free-free transitions, or Bremsstrahlung, in electron–proton collisions,
(2)   Infra-red emission by excited hydrogen molecules.

A consideration of the cross section for Bremsstrahlung shows that the process (1) is quite ineffective in reducing the temperature of the cloud....On the other hand, it can readily be shown that any appreciable percentage of hydrogen molecules must be highly effective in radiating the energy supplied by selective absorption. This can be seen by considering an assembly of hydrogen (protons and H_2^+) illuminated by a source of temperature T.

The discussion goes on to calculate the net heating per cm^3 from photoionization followed by recombination as the recombination rate times kT with a recombination cross section $\sigma = 4 \times 10^{-17}/T'$, where T' is the gas temperature. They then consider cooling by H_2 rotational and vibrational transitions:

Consider now the excitation of a hydrogen molecule by collisions with electrons. Interchange of energy between the translatory energy of the electron and the rotational states of the hydrogen molecule will take place freely. But the interchange with the vibrational states is somewhat more intricate and the cross section for complete equipartition between the electron and the molecule may be roughly estimated as 10^{-2} times the kinetic cross section or...$\sigma' = 2 \times 10^{-18}$ cm^2.

The energy 'supplied by the electrons to the gas', or the energy loss per excitation, is taken as equal to kT' times the collisional excitation rate. After equating the net heating and cooling rates they show that the gas temperature T' must be small compared with T unless the ratio [of electrons to hydrogen molecules] is very large.

In the next paragraph we see recognition of the importance of collisional de-excitation to cooling, or rather the absence of collisional de-excitation, even for highly forbidden transitions:

The foregoing calculation assumes that between successive excitations of a particular molecule, the molecule has sufficient time to radiate away the energy acquired in the first excitation. In connexion with this point it may be noticed that the excitation involved here concerns only the lowest electronic state of the molecule. The corresponding transitions are forbidden ones, since the electric dipole moment of the hydrogen molecule in its lowest electronic state is zero. Accordingly, the required transitions are due to a quadrupole moment, and since the emission of the radiation takes place in the infra-red the probability of these transitions may be expected, by analogy with the atomic case, to be smaller than the probability of an allowed transition by a factor of about 10^{-8}. If, therefore, we take 10^{-6} sec as the lifetime for an allowed transition we may write 10^{-2} sec as the approximate lifetime for the forbidden transitions. This lifetime is very short compared with the average time between successive excitations of the molecule. This may easily be seen; for if we suppose that the number of electrons per cm^3 is about 10^3 and if we take their velocity as 5×10^7 cm per sec, the probability per molecule of exciting collisions per sec is

$$2 \times 10^{-18} \times 10^3 \times 5 \times 10^7 = 10^{-7}$$

(the factor 2×10^{-18} being the cross-section for this process), and the time between successive excitations of any molecule is accordingly of the order of 10^7 sec. Thus in spite of the fact that in the above calculation all the estimates made involve a considerable margin of safety, the lifetime is nevertheless only 10^{-5} of the interval between successive excitations.

This is the first realization in the literature of the importance of cooling by molecular hydrogen.

4.2.2   The lifetime of molecules of interstellar hydrogen

Hoyle and Lyttleton formulated the problem of the photoionization and photodissociation of H_2 in terms of gas flowing through the interstellar medium at about 5 km/s and encountering the ultraviolet radiation from an O or B star.

> ... for a hydrogen molecule to undergo ionization and dissociation
> which requires radiation of energy greater than 15.6 e.V., the molecule
> must approach the critical sphere surrounding some star. This sphere
> will in general be largest for O stars, but the density of O stars in
> space is so much smaller than for B stars that the B stars may
> altogether have more effect.

They then go on to show that the lifetime of molecular hydrogen in interstellar space under these conditions would be extremely long, and that any molecules which formed would remain until they actually entered the HII region. Even in that case they would have time to do some cooling.

4.2.3   Existence of interstellar molecules

Near the end of this paper is a section addressing the question of the formation of interstellar molecules. This section was written and submitted for publication shortly before the first observations of interstellar CH lines in stellar spectra, a point that the authors comment on at the end of the section. Again they are considering a pure hydrogen gas. They clearly recognize that normal three-body reactions are not possible in the interstellar medium, and that some special two-body formation process is required. The mechanism they suggest, two-body radiative association of hydrogen atoms in a relatively dense cloud, is not the correct mechanism even in the absence of dust. They argue that given enough time H_2 will form. What is most interesting and valid is the argument that in a gas with no heavy elements some hydrogen molecules must form to enable cooling before star formation can take place:

> It is now appropriate to investigate shortly the question of how
> interstellar molecules may come to be formed. In this connexion it
> may be noticed that in order to calculate the time required for a cloud
> of known density and temperature to form a comparable density of
> hydrogen molecules, some estimate has at the present time to be
> made for the cross-section for the formation of H_2 by two body
> collisions ... If the stars are regarded as forming before the molecules
> the cosmical cloud may be taken to have a temperature of some
> 10,000 K in accordance with Eddington's estimate. On the other hand

it seems probable that it is by accretion that the stars form in the first
place, and in this case it would appear that the molecules must form
before the condensation of the stars can begin.

The cooling problem addressed in 1940 by Hoyle and Lyttleton in the context
of accretion is analogous to the cooling problem in the early Universe where
the gas contains no heavy elements and even a small amount of molecular
hydrogen is critical for cooling and formation of the first generation of stars.
Their suggestion that some H_2 will form by pure gas-phase processes is correct,
although the specific mechanism, two-body association, is wrong.

   Another aspect of this work, which is not correct in detail, is the assumption
that molecular hydrogen can be photodissociated only by being exposed to very
hard ultraviolet photons with energy greater than 15.6 eV inside an HII region.
However, H_2 can be photodissociated by absorption at energies greater than
only 11.1 eV in the Lyman bands followed by emission into the continuum in a
process I described in 1966 (see Field 1966). Thus H_2 formed at a low rate would
be subject to photodissociation even outside an HII region. However, H_2 formed
on grains (in a metal-rich environment) at a high rate will be self shielded by
line opacity in all clouds with a hydrogen density greater than about 100 cm^{-3},
resulting in clouds that are almost completely molecular hydrogen (Solomon
and Wickramasinghe 1969).

4.3   The fragmentation of gas clouds into galaxies and stars

   This paper (Hoyle 1953) presents a theory for the formation of galaxies
and stars in a grand scheme of isothermal gravitational contraction, followed by
fragmentation, then further contraction and fragmentation into smaller units
in a repeating series which stops when the cooling becomes inefficient owing
to the high opacity of the fragments. At this stage the process ends since the
cooling time becomes much greater than the collapse time. Among the questions
addressed are:

   Why are the masses of galaxies confined mainly in the range 3×10^9
         to $3 \times 10^{11} M_\odot$?
   Why is the typical mass of a Type II star of order $M_\odot$?
   Why do Type II stars apparently form almost simultaneously with the
         origin of a galaxy?

Hoyle also speculates on the formation of Type I stars near the end of the
paper, where there is a prescient discussion of processes that may inhibit star
formation and enable a spiral galaxy disc to maintain a large gas content.

Here I concentrate on the star formation aspects of this paper, although the general process is applied to galaxies and stars. The model of 'hierarchical fragmentation' is based on thermal stability of the gas at a temperature of 10 000 K and the gravitational instability of a fragment with more than the Jeans mass. Using the notation of the time the collapse criterion was given as

$$\frac{GM}{V^{1/3}} > 5\Re T,$$

where V is the volume of the cloud. The Jeans mass expressed in current notation is approximately

$$M \approx (\pi k / G m_p \mu)^{3/2} T^{3/2} \rho^{-1/2}.$$

The key to the mechanism is the decrease in collapse time for each successive generation of fragmentation, which leads to a runaway process as the collapse time $t \approx (G\rho)^{-1/2}$ decreases with decreasing Jeans mass resulting from isothermal contraction.

The first part of the paper is a detailed discussion of the temperature of a pure hydrogen gas heated by turbulence with collisional ionization. The turbulence is assumed to be at velocities >10 km/s, appropriate to the formation of galaxies and Type (Population) II stars. After considering the cooling curve and ionization equilibrium he concludes that there will be two stable temperature regimes: '*either the temperature lies in the range 10,000°–25,000° or the temperature must be high, of the order of $3 \times 10^{5°}$ or more.*' Fragmentation is then considered in detail for the 10 000 K regime, leading to the formation of ever smaller sub-units until stellar masses are reached. The process is described in the paper beginning with a galactic mass:

MODEL FOR THE HIERARCHY STRUCTURE

Step 1.—A spherical galactic condensation of mass $3.6 \times 10^9 \, M_\odot = M_0$, say, temperature 10^4 K, initial density $\rho_0 = 10^{-27}$ gm/cm^3, initial radius $R_0 = 1.2 \times 10^{23}$ cm, compacts by a factor $k^{2/3}$ and then divides into k equal masses, each of radius R_0/k, density $\rho_0 k^2$, and temperature 10^4 K.

It will be noticed that the obvious requirements are satisfied by these assumptions. Thus each of the subunits satisfies the contraction condition. Moreover, with their assumed radii the subunits can be fitted into the contracted volume of the galaxy itself. Remembering that in the first shrinkage the galaxy decreases in dimensions by a factor of about 3, it follows that we must put $k^{2/3} \sim 3$, which gives $k \sim 5$.

Step 2.–The subunits of step 1 themselves contract by a factor of $k^{2/3}$ and then divide into k equal fragments, each of radius R_0/k^2, density $\rho_0 k^4$, and temperature 10,000 K. Once again these smaller fragments satisfy the contraction condition.

Step 3.–The subunits of step 1 themselves contract by a factor of $k^{2/3}$ and then divide into k equal still smaller fragments, each of radius R_0/k^3, density $\rho_0 k^6$, and temperature 10,000 K. Once again these still smaller fragments satisfy the contraction condition.

And so on for further steps.–Now, since the time scale for condensation at each step may be expected to follow the inverse square root of the density, it follows that the ratio of the time required for an infinity of such steps to the time required for just the first step is given by

$$1 + \frac{1}{k} + \frac{1}{k^2} + \frac{1}{k^3} + \cdots = \frac{k}{k-1},$$

which for $k = 5$ gives a ratio close to unity. Thus we see that the main expenditure of time lies in the first step and that further steps require comparatively little time. It is indeed this rapid convergence of the steps that provides the strongest reason for the opinion that dissipation arises from the development of the sort of hierarchy structure postulated above.

The importance of the isothermal nature of the process is stressed, and an explanation is given for the inability of the pure hydrogen gas to cool below 10 000 K:

> Now since the time scale for each successive step decreases as $1/k$, it follows that the rate at which thermal energy is made available per unit mass increases by a factor of k at each step. On the other hand, the rate of radiation per unit mass increases by a factor k at each step. Accordingly, radiation is favoured by the changes that take place as fragmentation proceeds. There is, therefore, a tendency for the temperature to fall, *but the temperature cannot fall much below 10,000 K, since otherwise the hydrogen would become insufficiently ionized for effective radiation to occur.* This conclusion is subject to the implicit assumption that the fragments remain optically thin to radiation. The entire nature of the condensation process is altered if the fragments become optically thick to the whole of the radiation spectrum of the hydrogen. A discussion of this issue forms the main topic of the following section.

The paper goes on to consider the conditions which might terminate the process and result in the fragmentation of a galaxy into Type II stars:

> For a transition to the adiabatic condition, we require L to become less than the rate at which gravitational energy is being released in our problem. The gravitational energy released per fragment in the nth step of the hierarchy process is of the order of
>
> $$\frac{GM_0^2}{R_0 k^n}(k^{2/3} - 1),$$
>
> the release occupying a time of order $k^{-n}(G\rho_0)^{-1/2}$. In this time the energy radiated is $L k^{-n}(G\rho_0)^{-1/2}$. Hence a transition from the isothermal to the adiabatic condition begins to occur at the stage where
>
> $$L = \frac{G^{3/2}M_0^2\rho_0^{1/2}}{R_0},$$

The opacity of the gas is the limiting factor that determines the luminosity and the mass of the final fragments. The remainder of the treatment of the formation of Type II stars consists of a discussion of the opacity and the strong effect of a small admixture of metals on the opacity coefficient. In particular, for temperatures of about 3000–5000 K the electrons supplied by metals form H^-, which is the dominant opacity source. At lower temperatures heavy molecules may raise the opacity further. Using the negative hydrogen ion as the main source of opacity with a metal abundance of about 5×10^{-6} in the predominantly neutral gas, Hoyle arrives at a limiting mass for the fragments of between 0.3 and 1.5 $M_\odot$.

In the 1953 paper there is no mention of the importance of cooling by H_2 in a pure hydrogen gas, in contrast to the 1940 paper. While H_2 cooling was the basis for a paper 13 years earlier, this paper considers only atomic hydrogen modified by the addition of some metals. I can only guess at the reason for this. In between these two papers the 21-cm line of HI was discovered and the Galaxy was found to be full of HI. This may have led Fred, along with most other astronomers, to forget about hydrogen molecules. I do know that when in 1968–9 I discussed with him my own work involving the abundance and self-shielding of H_2 and in 1974 evidence for huge quantities of H_2 in dark clouds from CO observations and particularly our observations of the Milky Way molecular ring, he was very interested. He remarked that the 21-cm astronomers including Oort had been very confident that they had found almost all of the hydrogen.

There were many papers in the 1960s and 1970s that improved the Hoyle fragmentation picture by including the influence of other cooling agents, e.g. H_2, a more accurate treatment of heavy elements, molecules and dust, in order to arrive at a more accurate picture. However, Rees (1976) showed that the final mass of the fragmentation process is insensitive to the final temperature or the detailed opacity and can be expressed as a fraction of the Chandrasekhar mass, which explains why fragmentation results primarily in objects of stellar mass and not of galactic or planetary mass.

The last section of the fragmentation paper is in some ways the most modern. Hoyle speculates on the formation of Type I stars, that is stars forming from a gas with high metallicity and substantial dust in a spiral galaxy disc. This cloud is denser and much cooler than the huge low-density cloud that starts the fragmentation in a collapsing galaxy. As a result, the timescale for fragmentation is much smaller, which presents a serious problem. Hoyle takes as a starting point for the fragmentation leading to the formation of disc stars an interstellar cloud of mass $10^4 \, M_\odot$, $T = 50$ K and density $\rho_0 = 10^{-23}$ g/cm^3:

> ...comparable with the masses of the larger interstellar clouds in the galaxy. The fragmentation time for such a cloud would be less than 4×10^7 years...On this basis we should at first sight expect the residuum of gas to become entirely condensed into stars in only 4×10^7 years, reckoned from the time that its density rises high enough for tidal shearing to be unimportant. But such an exhaustion of the remaining gas certainly has not occurred in the galaxy or in other spirals. The question is: Why?

There is a clear realization that in the Milky Way disc and other spirals there must be a process at work that limits the efficiency or regulates star formation. Unimpeded star formation in the disc will lead to what would now be called a starburst. Although it is fashionable to discuss star-formation-induced star formation or supernova-induced star formation, the more important process is the inhibition of star formation. In normal spiral galaxies, star formation on the scale of the whole disc is a slow and inefficient process, although it may be occasionally efficient on very small scales. Here are Fred Hoyle's thoughts on this problem, which still sound current 50 years later, although the problem is still far from a quantitative solution. He makes three 'tentative' suggestions; the last two are the most interesting:

> b)...If very luminous stars of high temperature are formed – and there is, of course, observational evidence that this is so – a considerable heating must occur in the gas. [At this point there is a footnote to the forthcoming paper by Oort and Spitzer, and an acknowledgement to Lyman Spitzer for a private communication on the subject of heating.]

It is readily possible for a few highly luminous O stars to heat a gas cloud of Mass $10^4\odot$ to such a degree that the cloud is forced to expand under the action of gas-pressure forces – not of radiation pressure. Thus the following situation may well arise: When there are no luminous high-temperature stars, condensation occurs, and O stars are formed. But as soon as O stars are formed, the clouds are expanded and condensation ceases. After the bright stars become exhausted, the clouds then re-form and more stars condense. On this picture a sort of cyclic process is set up – clouds condense; stars form, including O stars; clouds blow up, and star formation ceases; O stars die; clouds re-form; more stars condense, including O stars; etc. The interstellar clouds of small mass would represent bits of clouds left over from the blowing-up process.

c) It has in recent years been thought that the galaxy may possess a substantial magnetic field. It has been suggested that the magnetic field owes its origin to the general motions of ionized interstellar clouds. But this suggestion has the disadvantage, from the point of view of explaining the polarization properties of the interstellar medium, that the magnetic vector cannot apparently maintain its direction over sufficiently great distances. This difficulty is much alleviated if it be supposed that the magnetic field is built up, not in the motions of the interstellar clouds, but by the motions of the original gas cloud out of which the galaxy originally formed. In this case the magnetic vector would also alter in direction from point to point, but alignment might be expected over far greater distances than on the earlier view. Moreover, the concentration of the residuum of gas to a disklike shape would tend to align such a magnetic field parallel to the galactic plane, in accordance with the requirement of Davis and Greenstein in their theory of the polarization properties of the medium.

Now it may occur, after a sufficient fraction of the gas in the disk has become condensed into stars, that the self-gravitation of the remaining gas is insufficient to overcome the magnetic pressure, in which case there is no tendency for large-scale condensation into stars. In exceptional regions, self-gravitation may indeed overcome the magnetic pressure – for example, in the Orion nebula – and extensive condensation may occur there. In this way, it might be possible to explain why a gaseous disk has in the main managed to survive in the galaxy and in other spirals for several billion years.

4.4   The composition of interstellar grains

Fred Hoyle produced a large body of work in conjunction with Chandra Wickramasinghe addressing the question of the composition of interstellar grains. Their first paper in 1962 challenged the prevailing view that grains were composed primarily of ice and instead were composed of graphite formed in carbon stars and ejected into the interstellar medium by radiation pressure. In 1970 the idea of a stellar origin of grains was extended to include supernovae (with D. Clayton). It is now generally accepted that interstellar grains originate in stars. In a series of papers in the mid 1970s Hoyle and Wickramasinghe suggested that complex organic compounds were a major component of grains. This included the then radical idea that heterocyclic hydrocarbons could explain features in the interstellar dust spectrum. During the 1980s they suggested that interstellar grains were not only organic, but biological material, including bacteria. In all of this work they used the spectrum of interstellar dust (absorption and emission) from the ultraviolet to the mid infrared in order to identify the constituents of grains. I shall briefly discuss some of the highlights from their early work (up until about 1977) below. A much more complete review including the later work is presented by Chandra Wickramasinghe elsewhere in these proceedings.

4.4.1   Graphite particles as interstellar grains

In this important paper Hoyle and Wickramasinghe both examine the observational evidence for graphite and develop a theoretical model for the formation of graphite in cool carbon-rich stellar atmospheres and the consequent expulsion into the ISM. Graphite came to be regarded as an important constituent of the ISM for the next decade. Hoyle and Wickramasinghe changed their view on graphite in favour of complex hydrocarbons in 1977–8, based on the new ultraviolet observations of the extinction curve near 2200 Å.

Their convincing arguments advanced in favour of graphite were developed in detail; they can be briefly summarized as follows:

(1)   The high average extinction requires that grains be composed primarily of the abundant elements carbon and oxygen. Water ice is possible, but a carbon compound such as graphite should be considered.

(2)   The theoretical extinction curve for graphite fits the observed extinction (see Figures 4.1 and 4.2).

(3)   Graphite has a sufficiently high albedo to explain reflection nebula.

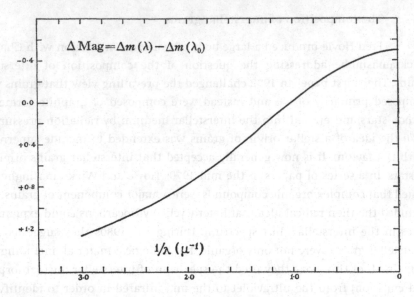

Figure 4.1 Theoretical curve of interstellar reddening (for graphite). [Originally Figure 1 in Hoyle and Wickramasinghe 1962.]

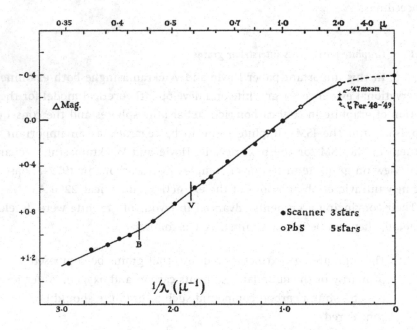

Figure 4.2 Normalized interstellar reddening curve derived from photoelectric scanner observations, and from infrared filter observations with a lead sulphide photoconductive cell. [Originally Figure 2 in Hoyle and Wickramasinghe 1962.]

(4)   Graphite cannot form in the ISM owing to the very low growth rate at low density. Stars forming graphite must have an excess concentration of C over O since CO will take up almost all of the less abundant constituent. Graphite can form in pulsating late-type carbon stars during the coolest part of the pulsation cycle when the partial pressure of gaseous carbon exceeds the vapour pressure of bulk graphite by a small margin.

(5)   Radiation pressure in the outermost layers can eject the grains into the ISM.

The summary of the paper in the abstract shows the full range of factors considered in favour of graphite:

> The interstellar reddening curve predicted theoretically for small graphite flakes is in remarkable agreement with the observed reddening law, suggesting that the interstellar grains may be graphite and not ice. This possibility is not in contradiction with the high albedos of reflection nebulae at photographic wave-lengths, provided the particles have sizes of order 10^{-5} cm.
>
> The origin of graphite flakes at the surfaces of cool carbon stars is considered, about 10^4 N stars in the galaxy being sufficient to produce the required density of interstellar grains in a time of 3×10^9 years.
>
> Grains tend to be formed in the pulsation cycle of an N star at temperatures < 2700 K. The grains have an important effect on the photospheric opacity, causing the photospheric density to decrease very markedly as the temperature falls towards 2000 K. It is this fall of density that allows the grains to be repelled outwards by radiation pressure and to leave the star altogether in spite of the frictional resistance of the photospheric gases. The grains do not evaporate as they leave the atmosphere of the star...
>
> ...Graphite is highly refractory and would not evaporate in H II regions. Graphite chemisorbs hydrogen and is therefore an effective catalyst in the production of interstellar H_2. Indeed graphite grains would be highly efficient in the production of interstellar molecules in general...

4.4.2   *Organic compounds as interstellar grains*

In the 1970s new spectroscopic data from space based ultraviolet and ground-based infrared observations led to a large increase in the data available for the identification of interstellar grains. In a series of papers in 1976–7 Hoyle

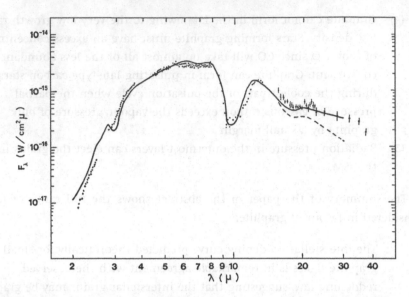

Figure 4.3 Points represent flux data in the 2–40 μm region for OH 26.5 + 0.6
(Forrest *et al.* 1978). Solid curve is the best-fitting-normalized flux from a
polysaccharide grain model with $T = 430\,$K, $\alpha = 2.6$. (Dashed curve omits
contribution to longwave flux from cooler cloud – see text.) [Originally
Figure 2 in Hoyle and Wickramasinghe 1977a.]

and Wickramasinghe shifted their attention to organic compounds and polymers
as the source of ultraviolet and infrared features. For example, they compared
the infrared emission expected from polysaccharide grains with the observed
spectra from a wide variety of sources (Hoyle and Wickramasinghe 1977b). For
some sources such as OH 26.5 + 0.6 they showed a comparison between the
observed spectrum from 2 μm to 30 μm and the expected emission from an
optically thin source of polysaccharides at 400 K (Hoyle and Wickramasinghe
1977a). See Figure 4.3.

The goodness of fit over a wide spectral range is taken as evidence of a correct identification, although they leave open the possibility that other organic
compounds are important:

> While it is true that the absorption peaks at approximately 3, 3.4 and
> 10 μm are common to many organic compounds we could find no
> other condensed molecular solid with a spectrum to match a wide
> range of astronomical data.

At first Hoyle and Wickramasinghe suggested that the polysaccharides might be
produced from gaseous formaldehyde in interstellar space, but a few months
later they suggested (Hoyle and Wickramasinghe 1977d) that polysaccharides

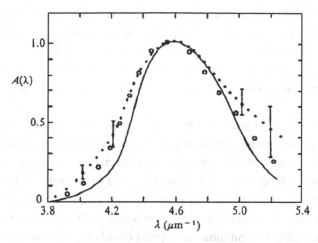

Figure 4.4 Normalized average molar absorptivity for $C_8H_6N_2$ isomers (solid curve) compared with interstellar extinction data in the waveband $3.8 \ \mu m^{-1} < \lambda^{-1} < 5.4 \ \mu m^{-1}$. Normalization is to 0.0 at $\lambda^{-1} = 3.8 \ \mu m^{-1}$, 1.0 at $\lambda^{-1} = 4.55 \ \mu m^{-1}$. Vertical bars give indication of spread of astronomical data. The dotted curve is the mean extinction curve of Bless and Savage (1972). Open circles give mean extinction $\langle E(\lambda - V)/E(B - V)\rangle$ relative to extinction data for θ-Orionis, and are normalized as above. [Originally Figure 1 in Hoyle and Wickramasinghe 1977c.]

could form in an outflow from a very young O star with a huge mass loss rate and an optically thick cooler envelope, based on an outflow model for strong infrared sources without HII regions (Hoyle *et al.* 1973).

At about the same time they realized that graphite by itself could not produce the shape of the important ultraviolet feature in interstellar extinction at 2200 Å. Rather than pure graphite as the source of the ultraviolet feature they suggested hydrocarbons in the form of nitrogenated heterocyclic hydrocarbons (Hoyle and Wickramasinghe 1977c). They compared the mean observed extinction near 2200 Å with measured absorptivity for the hydrocarbons, showing that the peak wavelengths agreed very well and the shape was close, although not perfect (see Figure 4.4). Hoyle and Wickramasinghe also commented that the observed strength of this feature would require only about 10% of interstellar C and N in the form of the hydrocarbons.

Although these specific hydrocarbons may not be the dominant form in the interstellar medium, it is now generally accepted that some type of hydrocarbon, possibly polycyclic aromatic hydrocarbons, or PAHs, are an important component of interstellar matter.

As is well known, Hoyle and Wickramasinghe went on to suggest that even more complex compounds are present in interstellar matter and in comets, including life itself in the form of bacteria. Fred Hoyle firmly believed that life could not have originated on Earth. He felt that the probability of life forming from non-living material was simply too small to happen on Earth. Although it is generally thought that he adopted his views on the effect of interstellar matter on the Earth late in his career, at least outside his science fiction, this is not the case.

In 1939, when Fred was 24 years old, he wrote a paper with R. Lyttleton titled 'The effect of interstellar matter on climatic variation' (Hoyle and Lyttleton 1939). This was an application of their accretion mechanism, which suggested that when the Sun passed through a dense interstellar cloud the accretion rate would be sufficient to temporarily increase the solar luminosity and warm the Earth. Ice ages would correspond to periods of no interaction with interstellar clouds. Accretion is not that effective and most interstellar clouds are not that dense, but this paper shows the breadth of his thinking throughout his career. Although not correct on the origin of the Earth's climate variation, he was certainly right a decade later in developing a theory for the origin of the elements on the Earth, one of the great scientific accomplishments of his time.

References

ATKINSON, R. D'E. 1940 *Proc. Camb. Phil. Soc.*, **36**, 314

BLESS, R. C. & SAVAGE, B. D. 1972 *ApJ*, **171**, 293

FIELD, G. B. 1966 *ARA&A*, **4**, 207

FORREST, W. J., HOUCK, J. R., McCARTHY, J. F. *et al.* 1978 *ApJ*, **219**, 114

HOYLE, F. 1953 *ApJ*, **118**, 513

HOYLE, F. & LYTTLETON, R. A. 1939 *Proc. Camb. Phil. Soc.*, **35**, 405

   1940 *Proc. Camb. Phil. Soc.*, **36**, 424

HOYLE, F., SOLOMON. P. M. & WOOLF, N. 1973 *ApJL*, **185**, 89

HOYLE, F. & WICKRAMASINGHE, N. C. 1962 *MNRAS*, **124**, 417

   1977a *MNRAS*, **181**, 51p

   1977b *Nature*, **268**, 610

   1977c *Nature*, **270**, 323

   1977d *Nature*, **270**, 701

REES, M. J. 1976 *MNRAS*, **176**, 483

SOLOMON, P. M. & WICKRAMASINGHE, N. C. 1969 *ApJ*, **158**, 449

5

Accretion

SIR HERMANN BONDI

Churchill College, University of Cambridge

The notion that the accretion of interstellar matter was an important factor in stellar evolution was presented forcefully by Hoyle and Lyttleton (1939, 1940). They pointed out that, with a star moving rectilinearly through a cloud of gas originally at rest, the orbits of the gas particles would intersect on a line behind the star, the 'accretion axis'. Collisions on this line would reduce to zero the transverse momentum of the particles, leaving a considerable proportion of the particles with insufficient kinetic energy to avoid falling into the star. It is readily seen that the capture cross section of the star by this process is of order $[GMV^{-2}]^2$, where G is the constant of gravitation, M the mass of the star and V its velocity relative to the cloud. In many circumstances this will greatly exceed the area presented by the body of the star to the cloud.

The collisions on the accretion axis will raise considerably the temperature of the gas, so creating a zone of high pressure, unless the cloud contains sufficient numbers of molecules and dust particles to radiate away the heat generated. This was a reasonable assumption in the light of observations beginning to be made at the time.

In 1942 I joined the radar section of the Admiralty Signals Establishment, where Fred Hoyle was already working. After various reorganizations and moves, Fred and I were not only working together in daytime on radar business, but often spent the evenings together. In those evening hours Fred enthused me (and Tommy Gold) about the problems of astrophysics. Before long I was working on the accretion question, particularly on the gas flow on and near the accretion axis. I was fortunate in that in my radar work on transmission values I had just

The Scientific Legacy of Fred Hoyle, ed. D. Gough.
Published by Cambridge University Press. © Cambridge University Press 2004.

become familiar with multi-stream states where there is more than one velocity vector at each point of the fluid. So I was able to make clear the complex flow patterns on and near the accretion axis. Along it there is a radial flow of high density in a narrow cylinder, the 'accretion column'. Its skin is the locus of the collisions that destroy the transverse velocity of the particles of the cloud. (For this work I was elected to a Junior Research Fellowship at Trinity College, Cambridge, in 1943.) However, it turned out that the steady-state equations did not determine the location of the crucial point of the accretion column within which the flow is towards the star and beyond which it is away from it. The position of this point determines the rate of accretion. Fred and I then integrated numerically (using the crude methods of the time) the equations for the time-dependent situation when a star enters a uniform cloud, moving at right angles to its originally plane boundary. We thus obtained a good order-of-magnitude estimate for the rate of accretion. We also established the formula for the drag exerted by the cloud on the star. All this work we published as a joint paper (Bondi and Hoyle 1944).

In the succeeding years we applied this notion of accretion to a variety of situations (stellar evolution, solar corona, etc.). Little of this work turned out to be of lasting importance.

However, in an idle moment a few years later I worked out what would happen if the star were at rest in the cloud. This spherically symmetric case seemed to me so simple and so artificial that I thought it might be useful only as a source of Tripos examination questions. Fortunately Lyttleton persuaded me to publish this work (Bondi 1952). Eventually the paper became a citation classic! It is also said to present the first astrophysical example of transsonic flow. But initially I thought of it as a highly academic exercise.

In the following year W. H. McCrea published a paper (McCrea 1953) that demonstrated that if circumstances were such as to result in appreciable accretion according to our 1944 paper, then the drag would very rapidly bring the star to rest in the cloud. Though Hoyle and I had written down the formula for the drag in that paper, we had failed to apply it! Accordingly, McCrea's paper shows that spherically symmetric accretion is effectively the only astronomically important case. So much for my Tripos questions!

In the nearly 50 years that have elapsed, the application of the notion of accretion has undergone great changes. Angular momentum, which we had neglected, is now central to the cases of accretion by protostellar and proto-galactic discs. Often the magnetic forces are also important. Perhaps accretion onto black holes is now the most important example of the relevance of our ideas.

References

BONDI, H. 1952 *MNRAS*, **112**, 195

BONDI, H. & HOYLE, F. 1944 *MNRAS*, **104**, 273

HOYLE, F. & LYTTLETON, R. A. 1939 *Proc. Camb. Phil. Soc.*, **35**, 405

1940 *Proc. Camb. Phil. Soc.*, **36**, 424

McCREA, W. H. 1953 *MNRAS*, **113**, 162

6

From dust to life

CHANDRA WICKRAMASINGHE

Cardiff Centre for Astrobiology, Cardiff University

The same stream of life that runs through my veins night and day runs through the Universe and dances in rhythmic measures. It is the same life that shoots in joy through the dust of the Earth in numberless blades of grass and breaks into tumultuous waves of leaves and flowers...

Rabindranath Tagore (1861–1941)

After initially challenging the dirty-ice theory of interstellar grains, Fred Hoyle and the present author proposed carbon (graphite) grains, mixtures of refractory grains, organic polymers, biochemicals and finally bacterial grains as models of interstellar dust. The present contribution summarizes this trend and reviews the main arguments supporting a modern version of panspermia.

6.1   Introduction

Dust abounds in the Universe – in interstellar space, in intergalactic space, in external galaxies, in planetary systems, in comets and on planets like the Earth. The many conspicuous dark lanes and striations seen on a clear night against the backdrop of the Milky Way are gigantic clouds of cosmic dust amounting in total mass to about one per cent of the material that lies between the stars. New stars, including planetary systems, condense from such clouds of interstellar dust, so the role of dust in astrophysics is by no means trivial.

The nature of interstellar dust has never ceased to spark fierce controversy and debate amongst astronomers (Wickramasinghe 1967). This situation has not

The Scientific Legacy of Fred Hoyle, ed. D. Gough.
Published by Cambridge University Press. © Cambridge University Press 2004.

significantly changed for nearly a century. In the 1930s the fashionable model of interstellar dust involved iron grains with sizes of a few hundredths of a micrometre, a model proposed by C. Schalen by analogy with the composition of iron micrometeoroids (Schalen 1939). Then in the mid 1940s the emphasis shifted to volatile grains of an icy composition, broadly similar to the particles that populate the cumulus clouds of the Earth's atmosphere. H. C. van de Hulst (1946) had argued that such grains would condense in interstellar clouds, and J. H. Oort and van de Hulst (1946) developed this theory in considerable detail. By the close of the 1950s the dirty-ice-grain model was one of the holy grails of astronomy. Astronomers had by then ceased to worry about how these grains might be formed, but concerned themselves only with the corrections that were required in order to take account of dust absorption and scattering for estimating the distances and temperatures of stars.

It was at this stage that Fred Hoyle introduced me to this topic, essentially inviting me to reopen the debate as to the composition and formation of grains. Fred was convinced at the outset that the prevailing point of view had to be wrong. His reason was connected with the problem of nucleating solid particles at the densities that prevailed in interstellar clouds, densities of less than a million atoms per cubic centimetre even in the denser clouds. If supersaturated clouds of water vapour do not lead to the production of ice crystals or rain drops in the troposphere without the introduction of condensation nuclei, what chance could there be for nucleation of ice in interstellar clouds? Fred's conjecture on this matter was easily checked by appealing to well documented results of homogeneous nucleation theory. If grain nuclei are absent, solid particles do not form under interstellar conditions, and one is therefore forced to consider astrophysical venues of very much higher density for the nucleation of grains. Such venues include atmospheres of cool stars (Hoyle and Wickramasinghe 1962) protoplanetary discs (Hoyle and Wickramasinghe 1968), and supernova ejecta (Hoyle and Wickramasinghe 1970), venues that we considered in succession as possible sites of grain formation.

6.2   Graphite grain theory

Carbon stars where the photospheric C/O ratio exceeds unity seemed a good bet for the condensation of carbon grains. Whilst oxygen would bond stably to carbon as CO (11.2 eV), the excess photospheric carbon would be available for condensation as solid carbon. Carbon stars commended themselves at the outset, particularly in view of the well documented behaviour of R Cor Bor, which varied in brightness by several magnitudes as it came to be shrouded sporadically in a shell of obscuring dust.

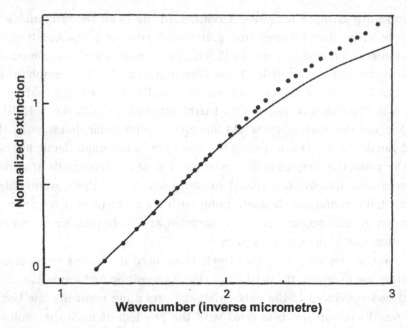

Figure 6.1 Nandy's (1964) observations of the average visual extinction law (points) compared with the standard Oort-van de Hulst model curve # 15 for dirty ice grains.

Our investigations were initially confined to the N stars, which are irregular variable carbon stars where the photospheric temperature varies typically in the range 1800–2500 K during a pulsation cycle lasting ∼100 days. Thermodynamic equilibrium calculations together with nucleation theory showed clearly that graphite particles would nucleate and grow near minimum phase (Hoyle and Wickramasinghe 1962; Donn *et al.* 1968). When the grains (spheres or flakes) grew from saturated gas to diameters of hundredths of a micrometre, radiation pressure from the parent star would exceed gravity and the particles (either flakes or spheres) would be expelled at high speed into interstellar space. We next argued that grains are selectively stopped in the denser interstellar clouds where they give rise to the well known extinction of starlight.

In the early 1960s observational data on the optical properties of interstellar dust were limited mostly to the optical waveband. The extinction at $\lambda V = 5500$ Å in directions close to the plane of the Galaxy in the solar vicinity amounted to ∼1.8–2 mag per kpc, and the wavelength dependence of the extinction gave a $\Delta m(\lambda)$ approximately proportional to inverse wavelength over the wavelength range 8000–3300 Å, as shown by the points displayed in Figure 6.1. There was also an observation of a few per cent interstellar polarization with a flat maximum near 5000 Å, and data on the scattering of light in reflection nebulae which required elongated grains to have a high visual albedo (exceeding 0.5) at the same wavelength (Wickramasinghe 1967). All the available data were roughly consistent with the ice-grain model of van de Hulst, as far as the optical data were concerned, but higher-resolution data even in the early 1960s obtained by K. Nandy (1964) showed some noteworthy discrepancies, as seen for instance in Figure 6.1.

The fit to the data in Figure 6.1 could in fact be improved if one used size distribution functions for grains other than the one used by Oort and van de Hulst (1946), but this procedure already exposed a serious shortcoming of the ice-grain model (Wickramasinghe 1967). Any reasonable model of the grains had to explain why the extinction law (points in Figure 6.1) remained invariant over the optical waveband in whatever direction one cared to look. On the basis of the ice model this had to be explained as a grain-size effect: the size distribution of grains would have to possess a sharply tuned mean effective radius that remained invariant to within a small percentage of $a = 0.2\,\mu m$ in all directions close to the Galactic plane. For grains that were supposed to condense from the gas phase under conditions of widely ranging density, this constraint was artificial and unsatisfactory.

When we first proposed the graphite grain theory, data on the optical constants of graphite were sparse. From the available data at only a few wavelengths we were able to model the optical constants on the assumption of constant conductivity, and hence use Mie theory to calculate the extinction properties of our model over the wavenumber range 1–3 inverse micrometres (Hoyle and Wickramasinghe 1962). Our first calculation for graphite particles yielded an excellent agreement with the observed visual extinction law provided particle radii were all less than $0.05\,\mu m$. This feature that the extinction law was not rigorously tied to any fixed value of particle size or size parameter was an undoubted asset of the model.

This particular advantage for graphite was, however, short-lived, and disappeared when the first detailed laboratory measurements of optical constants became available (Taft and Phillipp 1965). The original advantage we claimed for graphite was now replaced by a stronger point that appeared in its favour. With the new data the extinction by graphite spheres showed a conspicuous peak in the ultraviolet, which for particles of radii $0.02\,\mu m$ occurred at a wavelength close to 2200 Å (Wickramasinghe and Guillaume 1965; Stecher and Donn 1965). This result is shown in Figure 6.2.

The peak in Figure 6.2 shifts slightly to shorter wavelengths for particles less than $0.02\,\mu m$, but this shift is within the error bars of the early data on ultraviolet extinction (Stecher 1965), so the radius constraint seemed irrelevant at this stage. At about the same time as these calculations were made, the

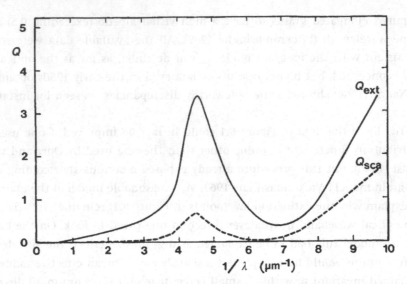

Figure 6.2 Extinction and scattering efficiencies of graphite spheres of radii
0.02 μm.

first measurements of the ultraviolet spectra of early-type stars revealed that
the interstellar extinction curve had a conspicuous peak at precisely the same
wavelength (Stecher 1965). Such a peak could not be explained with an ice-grain
model, so the new data provided an immediate impetus for graphite models.

With an evident victory for graphite over ice grains, Hoyle and I soon published a large volume of rigorous computations for models involving mixtures
of graphite particles with silicate grains and small iron grains. Just as we had
argued for graphite particles forming in carbon stars we soon showed that
mixtures of iron and silicate/silica grains would be expected to condense in
mass flows from oxygen-rich giant stars, the Mira variables in particular. Iron
and silicates as well as carbon particles were also shown to condense from
the ejecta of supernovae, particularly iron in the form of whiskers (Hoyle and
Wickramasinghe 1970, 1988).

With the discovery of an infrared emission peak in the dust envelopes of
M-giant stars (Woolf and Ney 1969; Knacke et al. 1969) the emphasis shifted
markedly away from volatile icy grains to refractory grain mixtures. This
trend was further consolidated because searches for 3.1- μm extinction features
in highly reddened stars had led consistently to negative results (Danielson,
et al. 1965). Although the precise role and quantity of silicates in interstellar
dust is still a matter of dispute, the 2175 Å absorption peak, which was later
found to be invariant with regard to its precise central wavelength, is still widely
attributed to graphite grains.

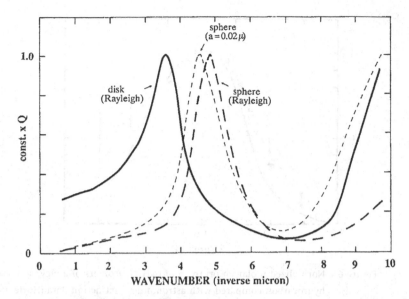

Figure 6.3 Calculation of extinction curve for graphite in the form of Rayleigh
scattering discs and spheres, compared with that for a sphere of radius
0.02 μm (Wickramasinghe *et al.* 1992).

6.3   The emergence of organics

Despite being the originators of the theory of graphite grains and prime
movers in a major paradigm shift from volatile icy grains to refractory mixtures,
Hoyle and I became disillusioned with the graphite model at any early date. The
problem for us was the firm requirement for graphite spheres of radius precisely 0.02 μm. In the real world graphite is a highly anisotropic crystal and
condenses as either whiskers or flakes. Spherical particles of a specific radius
represented a travesty of common sense. A calculation carried out by Wickramasinghe *et al.* (1992) illustrating this point is shown in Figure 6.3. This indicated
to us most clearly that graphite grains from carbon stars cannot contribute significantly to the overall composition of grains.

Disillusioned with the general class of graphite-based models, Hoyle and I
began in the mid 1970s to explore an alternative explanation of the 2175 Å extinction feature in terms of bicyclic aromatic compounds (Hoyle and Wickramasinghe 1977b). The fit of the astronomical data to our first polyaromatic hydrocarbon model based on $C_8H_6N_2$ isomers is shown in Figure 6.4, the model
requiring a large fraction of the carbon in the interstellar medium to be tied up
in the form of such molecules. It should be stated that this was the first suggestion in the literature of polyaromatic hydrocarbons (PAHs) in interstellar space,
a model that has come increasingly into vogue in recent years. Such molecules

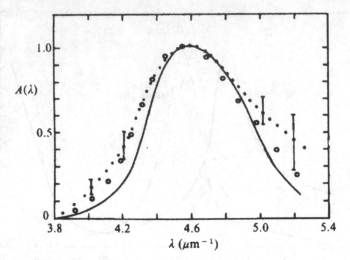

Figure 6.4 Normalized molar absorptivity of $C_8H_6N_2$ isomers of a bicyclic aromatic hydrocarbon compared with astronomical extinction data (Hoyle and Wickramasinghe 1977b).

are unlikely to form *in situ* from the gas phase under interstellar conditions, so we considered it likely that they are the result of degradation of more complex organic structures that condensed elsewhere.

The first explicit proposal for an organic composition of both interstellar and cometary dust was made in 1974/75 (Wickramasinghe 1974a,b; Vanysek and Wickramasinghe 1975). The original motivation was mainly the lack of a detailed agreement between the astronomical data for heated dust in the Trapezium Nebula and the calculated behaviour of real silicates over the infrared waveband 8–13 μm. Silicates were the favoured model for the dust composition in this region, but as shown in Figure 6.5 there were serious problems that had to be faced.

In the circumstance that no laboratory data for a silicate could fit the observations, the astronomical community pursued an unusual route. The observational data were themselves used to infer a fictional 'astronomical silicate', and thereafter the hypothetical 'silicate' was used as a standard for astronomical comparison. Hoyle and I argued strongly that this procedure represented an unacceptable inversion of logic (Hoyle and Wickramasinghe 1984). The lack of correspondence in Figure 6.5 was telling us something important, and that was a message that had to be deciphered, not ignored.

With cosmic abundance ratios implying that a small percentage of the mass of interstellar matter is composed of the elements C, N, O, and the interstellar extinction data demanding that the grains must make up a similar small

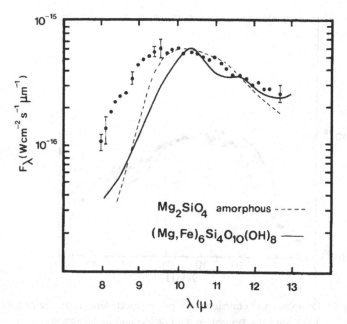

Figure 6.5 Data for emission by dust in the Trapezium Nebula (Forrest *et al.* 1975, 1976) that were discordant with both hydrated and amorphous silicates.

percentage of the interstellar material, one could immediately deduce that grains must *mostly* comprise the same elements, namely, C, N, O. These elements could occur in a variety of different combinations with hydrogen to make up solid matter. Since one such plausible composition, H_2O, was already ruled out by the astronomical data, combinations of C, N, O with H that form a large class of organic polymeric materials led to an alternative class of grain models. The discordance in Figure 6.5 was a strong incentive for this new class of model to be explored.

The question that I set out to answer in the summer of 1974 was whether the lack of adequate absorptivity/emissivity over the 8–9.5 μm waveband, which was an endemic problem for silicates, could be rectified with contributions from organic grains. Inspection of spectral atlases soon revealed that polymers involving C–O, C–N, C=C and C–O–C linkages do indeed offer the prospect of contributing over precisely this waveband, and a formaldehyde-based polymer first suggested itself in view of the ubiquitous occurrence of the gas-phase molecule H_2CO in interstellar space (Wickramasinghe 1974b). The first comparison of organic polymers with the Trapezium data is shown in Figure 6.6, where it is clear that the 8–9.5 μm waveband is particularly well served with an enhanced source of opacity.

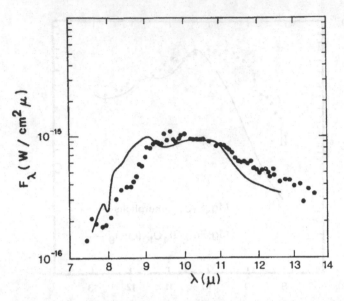

Figure 6.6 Comparison of emission from polyoxymethylene (POM) grains heated to 190 K with the Trapezium data (Wickramasinghe 1974b).

Encouraged by this discovery we proceeded to consider other organic polymers, moving progressively through various copolymers of formaldehyde through polysaccharides, sporopollenin to biochemicals. The fits to astronomical data improved systematically as we progressed along this sequence (Hoyle and Wickramasinghe 1977a, 1980a,b,c).

6.4   Absorptions at 3.4 μm

Whilst 10 and 20 μm absorptions could broadly be assigned either to silicates or organics (or a combination of the two), organic solids would be characterized uniquely by CH stretching absorptions at 3.3–3.4 μm. For most solids of organic composition, aliphatic CH stretching at 3.4 μm should in principle be observable. Could this be so in the astronomical case if, as we expected, the bulk of grain material is organic? An immediate difficulty arose because most organic polymers have a broad absorption profile centred around 3.4 μm, where the central mass absorption coefficient amounts to no more than $\sim$1000 cm^2 g^{-1}. Thus we require a path length of interstellar matter of $\sim$10 kpc to yield an optical depth at 3.4 μm of $\sim$0.1 mag. In the absence of any good data extending over such long galactic path lengths, Fred Hoyle and I spent nearly three years between 1976 and 1979 combing the published literature for vestiges of 3.4 μm features in a wide range of astronomical spectra (Hoyle and Wickramasinghe

1977a, 1980a,b,c). The evidence was clearly there, but it appeared in most cases so marginal to the eye that if one disliked the concept of organics for any reason it was possible to ignore it with impunity. But not for long.

A particular class of sources that attracted us involved dense clouds of interstellar dust for which absorptions at 3.1 μm due to water ice had in fact been observed. Hoyle and I were the first to draw attention to the fact that the 3.1 μm water ice band in such sources as the BN object (associated with protostellar clouds) always possessed a secondary weak 3.4 μm feature in the longwave wing of the ice band. It was weak enough to have gone unnoticed, but we insisted that in view of the low values of the mass absorption coefficient of organics near 3.4 μm (at least 30 times less than the absorption coefficient of water ice at 3.1 μm) the astronomical data implied an overwhelming preponderance of organic solid material. This appeared to be inevitable for grains within dense clouds, and as far as the data go there could be no question. The desire to sweep the data under the rug was so strong, however, that many bold refutations appeared in print. Thus Duley and Williams (1979) wrote: 'We conclude that no spectroscopic evidence exists to support the contention that much of interstellar dust consists of organic materials.' And Whittet (1979) similarly pronounced: 'It has become fashionable to discuss the possibility that interstellar grains may contain significant quantities of organic material. I draw attention here to new data which preclude the presence of organic solids as a significant component of the grains...' In the decades that followed, these pronouncements were of course all proved to be wrong.

6.5   Flirting with pre-biology

The conclusion that organic solid particles occur in profusion on a galactic scale could for long be isolated from the conjecture that this material was somehow associated with life. The connection first dawned on me in the autumn of 1975 after Vanysek and I had argued for organic polymers being the parent molecules of gases observed in cometary comae (Vanysek and Wickramasinghe 1975). Fred Hoyle was visiting Caltech and Cornell at the time, and I sought desperately to resume my contact with him. In a series of letters to him I made seemingly outrageous proposals about the possible connection between interstellar and cometary dust and biology. My correspondence spanned most of six months, and for a while Fred remained lukewarm to the idea of any connection. I think it was only when I had stressed to him the sheer quantity of organics that appear to be involved throughout the Galaxy that Fred made a speedy turn around. Our collaboration then promptly resumed, and early in 1976 we began toying with the idea of prebiotic evolution in interstellar clouds. The idea at this stage was that prebiological evolution proceeded in dense interstellar clouds to such an extent as to produce cell-like aggregates of organic matter. These particles, including organic molecules that are the complex building blocks of life, were then delivered to Earth by comets.

Our paper entitled 'Primitive grain clumps and organic compounds in carbonaceous chondrites' described a process by which organic grains would collide and stick together in interstellar space. A process of natural selection then led to the formation of stable cell-like clumps that acquired resistance to ultraviolet light. This paper, which was our first incursion into interstellar pre-biology, appeared in *Nature* (Hoyle and Wickramasinghe 1976). It should be said that such ideas were entertained as valid speculations for publication for the reason that they fell broadly within the prevailing paradigm for the origins and evolution of life. That situation was not destined to last.

6.6   The source GC-IRS7 and the move to biology

Although the presence of organic solid material in dense molecular clouds was not in dispute by the end of 1978, evidence for its widespread occurrence in the general interstellar medium was still a matter for argument. The best chance of detecting organic dust in the ISM was to detect a clear 3.4 μm signal in sources of infrared radiation located near the Galactic Centre 10 kpc away, for which the visual extinction would be some 30 magnitudes. Observations with a poor spectral resolution that were available before 1978 already provided tentative evidence of the kind we sought, but decisive data became available only in 1981. In 1981 D. A. Allen and D. T. Wickramasinghe used infrared equipment at the AAT to obtain a spectrum showing an unambiguous absorption feature in the Galactic Centre infrared source GC-IRS7 with a characteristic shape indicative of an assemblage of complex organic material in the grains. The extinction over a 10 kpc path length at the centre of the 3.4 μm band was as high as 0.3 mag (Allen and Wickramasinghe 1981).

Several months before this observation was made, my graduate student S. Al-Mufti had made the remarkable discovery that the 3.3–3.5 μm absorption of desiccated biological cells was substantially invariant, both with respect to cell type (prokaryotes or eukaryotes) and to ambient physical conditions, including temperatures up to ~700 K (Hoyle *et al.* 1982a,b; Al-Mufti 1984). The prediction then was that if grains were biological, GC-IRS7 must not only possess absorption in this waveband, but it must also show absorption with an optical depth that was a rather precisely determined function of wavelength. This is exactly

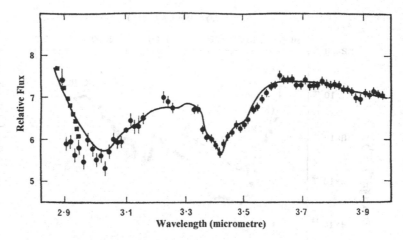

Figure 6.7 The first detailed observations of the Galactic Centre infrared source
GC-IRS7 (filled squares and circles) by Allen and Wickramasinghe (1981).
The solid curve is a prediction from earlier laboratory spectral data for
dehydrated bacteria (Al-Mufti 1984).

what was found when the first high-resolution spectral data for GC-IRS7 were
compared with the bacterial model. The modelling of the data was to all intents
and purposes unique, and the fit shown in Figure 6.7 implies that approximately
30 per cent of all the available carbon in the interstellar matter is tied up in
the form of material that is indistinguishable from bacteria.

We notice here that there is no evidence whatsoever in the data of an absorption at 3.1 μm due to water ice, which is consistent with other considerations
referred to earlier, and also provides independent testimony as to the superiority
of the observing site with regard to dryness of atmosphere, which is required
for reliable infrared observations.

The spectroscopic data obtained by Allen and Wickramasinghe (1981) have
of course been verified on numerous occasions by other observers (Okuda
et al. 1989, 1990; Pendleton *et al.* 1994). Recent attempts to measure the spectrum of GC-IRS7 have used more modern instruments, although not necessarily
at superior observing sites with regard to atmospheric dryness. The generally
favoured modern spectrum of GC-IRS7 appears to be one attributed to Pendleton
et al. (1994), which is reproduced as the points in Figure 6.8. We see immediately that this spectrum differs from the original spectrum of Allen and
Wickramasinghe (1981) (dashed line) to the extent of an excess absorption
over the 2.8–3.3 μm waveband that is generally consistent with the presence
of water ice. Our original conclusion concerning the E. coli   GC-IRS7 opacity
correspondence would remain valid provided we adopt one of the following

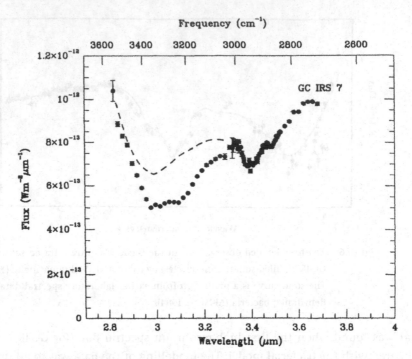

Figure 6.8 High-resolution data for GC-IRS7 (Pendleton *et al.* 1994) (points). Dashed curve is the average relative flux values from the data of Allen and Wickramasinghe (1981) and Okuda *et al.* (1989).

two procedures:

(i) subtract the excess absorption in this waveband, attributing it to spurious atmospheric water;

(ii) add a component of water ice to our proposed bacterial grains, an amount as little as 2% of the bacterial mass density being sufficient for this purpose.

Despite the modest nature of requirement (ii), I myself would prefer the former of these alternatives, option (i), and propose to adopt the relative flux curve of Figure 6.7 as having the correct overall shape, subject only to refinements of detail over the 3.4 μm band profile arising from improvements in astronomical spectroscopy.

The conclusion to be drawn from Figures 6.7 and 6.8 is that interstellar grains in the entire line of sight to the Galactic Centre have to possess infrared spectra indistinguishable from desiccated bacteria. All that the data minimally require is that organic functional groups exist in interstellar grains with a similar disposition to that which occurs in bacteria. So it would be legitimate for the critic to ask whether a non-biological grain that mimics a bacterium could be involved.

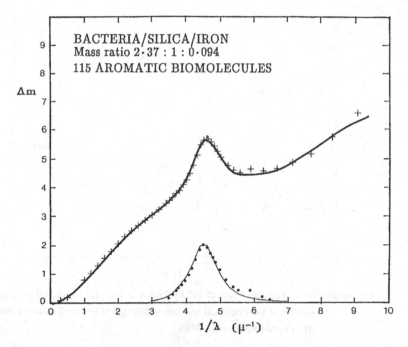

Figure 6.9 The filled circles (points) are excess interstellar absorption values over and above a scattering curve for hollow bacteria. Crosses are the mean interstellar extinction data. The heavy curve is calculated for hollow bacteria with an admixture of bioaromatic molecules and trace quantities of silica and iron in the form of submicrometre-sized grains. The thin line is the absorption profile for an ensemble of bioaromatic molecules. (Full references and credits in Hoyle and Wickramasinghe 1991.)

Our conviction that this is unlikely stemmed from many other considerations that we had brought to bear upon the problem. First, we had calculated the extinction behaviour of desiccated bacteria and found it to match the interstellar extinction law over the 0.5–3 inverse micrometre range to an amazing degree of precision without the need to fit any free parameters.

Our confidence grew further when we revisited our earlier suggestion that the ultraviolet extinction feature at 2175 Å was due to polycyclic aromatics. The degradation products of bacterial grains in space would include stable biological aromatic compounds and the spectrum of such an ensemble of molecules was simple enough to calculate from laboratory data.

The curve in Figure 6.9 shows the extinction behaviour for freeze-dried bacteria (with trace quantities of silicon and iron for fine tuning), and including 115 biological aromatic compounds. The points (crosses) are the astronomical observations for the total extinction of starlight (Hoyle and Wickramasinghe 1991)

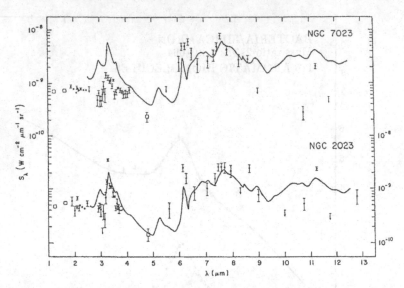

Figure 6.10 Integrated spectrum of 115 biological aromatics compared with the spectra of astronomical sources showing unidentified infrared bands (Wickramasinghe *et al.* 1990).

with the inset showing the contribution from the aromatics, compared to the 2175 Å absorption excess (Wickramasinghe *et al.* 1989, 1990). The data points in Figure 6.9 are a compilation of astronomical extinction data by Sapar and Kuusik (1978).

The biological ensemble of aromatic molecules responsible for ultraviolet extinction near 2175 Å would also have characteristic infrared absorptions, the behaviour of which is shown in Figure 6.10, compared with early data for two astronomical sources.

6.7   The 10 and 20 μm features in biological grains

We referred in Section 6.3 to the 8–40 μm spectrum of dust in the Trapezium nebula and the role of organics in explaining the deficiencies of silicate emission, particularly over the 8–10 μm waveband. We pointed out that many complex organic materials, including biopolymers, exhibit broad features arising from C–O, C=C, C–N, C–O–C bonds centred on wavelengths close to 10 and 20 μm. As our thoughts began to turn in the direction of cosmic biology, it occurred to us that there is a possible contribution from biogenically generated silica, as for instance is found in a class of algae known as diatoms, a class that appears to have made a sudden appearance on the Earth some 65 million years ago.

Al-Mufti, who was making laboratory measurements for all manner of possible candidate substances, obtained a mixed culture of diatoms taken from waters

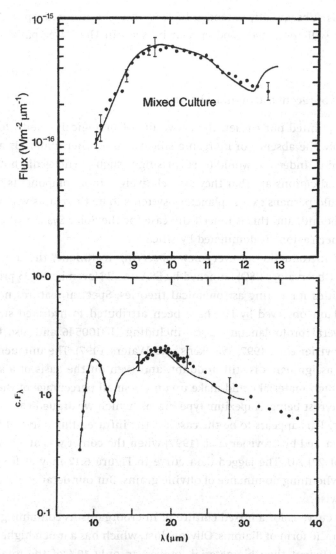

Figure 6.11 Points are data for the Trapezium nebula (Forrest *et al.* 1975, 1976;
Merril *et al.* 1976). The curves show calculated emission behaviour of
diatoms at 175 K.

of the River Taff (Hoyle *et al.* 1982c; Al-Mufti 1984). Here both 10 and 20 µm absorptions arise from a combination of biologically generated carbonaceous and siliceous material. Using laboratory data for this material together with a temperature of 175 K permits the expected emission of diatoms to be worked out at each wavelength λ. Thus the curve in the upper panel of Figure 6.11 shows the expected curve for diatoms. When this curve is compared with the observed points for Trapezium dust the agreement is seen to be impressive. And when the comparison was subsequently extended further into the infrared up to 40 µm, the agreement still remained good, as can be seen in the lower panel of Figure 6.11.

6.8   Trends of recent astronomical data

As we pointed out earlier, the above fits of biological systems to spectra do not imply the absence of inorganic silicates and other materials as constituents of grains. Indeed, it would be surprising if such particles did not exist, although the indications are that they are relatively minor components on the whole. One would perhaps expect planetary systems to be the places where mineral silicates abound, and this is indeed the case for the Solar System where the dust in the inner regions is dominated by silicates.

As with the introduction of every new observing technique, the use of the Infrared Space Observatory (ISO) launched by ESA on 17 November 1995 provided new opportunities for testing astronomical theories. Spectral features near 19, 24, 28, and 34 µm observed by ISO have been attributed to hydrated silicates, such as in several protoplanetary discs, including HD100546 and also Comet Hale–Bopp (Crovisier et al. 1997; Waelkens and Waters 1997). The uniqueness of some of these assignments is still in doubt, and even on the basis of a silicate identification such material could make up only a small percentage of the mass of the dust, the rest being Trapezium-type grains, which we argued earlier were largely organic. This appears to be the case for the infrared flux curve of Comet Hale–Bopp, obtained by Crovisier et al. (1997) when the comet was at a heliocentric distance of 2.9 AU. The jagged data curve in Figure 6.12 may at first sight imply an overwhelming dominance of olivine grains. But our detailed modelling showed otherwise.

The dashed curve is for a mixed culture of microorganisms containing about 20% by mass in the form of diatoms. Olivine dust, which has a much higher mass absorption coefficient than biomaterial, makes up only 10% of the total mass in this model (Wickramasinghe and Hoyle 1997, 1999). There are temperature parameters that have to be fitted in such a calculation, but here, as in many other similar instances, we find that our conclusions about the dominance of organics over minerals are valid.

The ISO has also produced an impressive set of data on infrared emission bands. Figure 6.13 shows a particularly interesting case where the spectrum of the colliding Antennae Galaxies at some 63 million light years distance has been observed to have infrared emissions that match the laboratory spectrum of anthracite (Guillois et al. 1999). Anthracite, being a product of biological (bacterial) degradation, is again indicative of biological particles, in this case at great distances from the Milky Way.

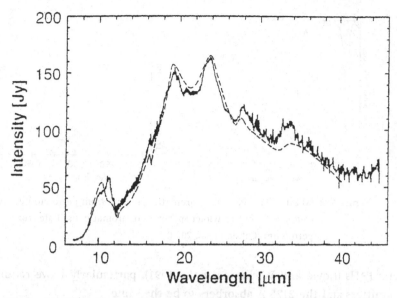

Figure 6.12 The spectrum of Comet Hale–Bopp (jagged curve) compared with computed flux from a mixture of olivine and biomaterial (diatoms + *E. coli*) (dashed curve), the olivine making up 10% of the total mass.

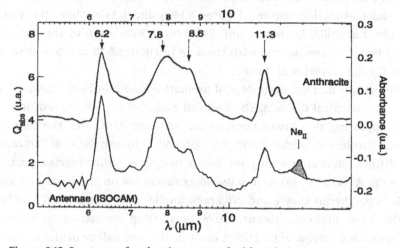

Figure 6.13 Spectrum of anthracite compared with emission spectrum of an entire galaxy (Guillois *et al.* 1999).

The distributions of unidentified infrared bands (UIBs) between 3.3 and 22 μm are almost identical in their wavelengths in very different emission sources, more-or-less irrespective of the ambient conditions. Whilst polyaromatic hydrocarbons (PAHs), presumed to form inorganically by some unknown process, are the favoured model for the UIBs, it has to be admitted that no really satisfactory agreement with the available astronomical data has been shown possible for

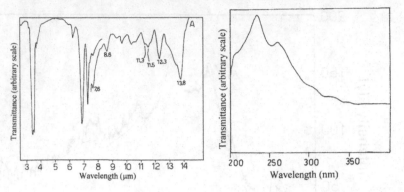

Figure 6.14 (a) Infrared spectrum of aromatic distillate from petroleum (Cataldo
        et al. 2002). (b) Ultraviolet spectrum of aromatic distillate from
        petroleum (Cataldo *et al.* 2002).

abiotic PAHs (Hoyle and Wickramasinghe 1991), particularly if we require the
UIB emitters and the 2175 Å absorbers to be the same.

More recently Cataldo, Keheyan and Heymann (2002) have shown that aromatic distillates of petroleum, another biological product, exhibit correspon-
dences with the astronomical diffuse infrared bands (UIBs) as well as the 2175 Å
ultraviolet extinction feature. Figures 6.14(a) and 6.14(b) show the results of
Cataldo *et al.* (2002) for petroleum distillates. A tabulation of the principal infrared bands for comparison with the data in Figures 6.13 and 6.14 is set out in
Table 6.1 (e.g. Gezari *et al.* 1993).

Biologically derived ensembles of aromatics have also been shown to match
other astronomical datasets. The so-called extended red emission of interstellar
dust, appearing as a broad fluorescence emission band over the range 5000–
7500 Å (Hoyle and Wickramasinghe 1996; Wickramasinghe *et al.* 2002a,b), and
the diffuse interstellar absorption bands in optical stellar spectra, particularly
the 4430 Å feature, also have possible explanations on the basis of molecules
such as porphyrins (Hoyle and Wickramasinghe 1979; Johnson 1971, 1972).

The most dramatic recent discovery relating to astronomical aromatic
molecules is a conspicuous 2175 Å band in the lens galaxy of the gravitational
lens SBS0909 + 532, which has a redshift of $z = 0.83$ (Motta *et al.* 2002). The extinction curve for this galaxy is reproduced in Figure 6.15(a), with the dashed
curve representing a scattering background attributed to hollow bacterial grains.

The excess absorption over and above this scattering background, normalized to unity at the peak, is plotted as the points with error bars in Figure
6.15(b). The curve in this figure shows the absorption of biological aromatic
molecules similarly normalized (Hoyle and Wickramasinghe 1989; Wickramasinghe *et al.* 2002b) (also see Figure 6.9).

Table 6.1 *Distribution of principal infrared emission wavelengths*

Principal UIBs: average of various sources including protoplanetary nebulae and Antennae Galaxies	Biological aromatic ensemble: 115 aromatics	Aromatic distillate extracted from petroleum	Anthracite
3.3	3.3		3.3
3.4	3.4	3.4	3.4
	3.6	3.5	
	5.28		5.2
6.21	6.21	6.2	6.2
			6.7
6.9	6.9	6.8	
7.2	7.2	7.2	7.2
7.7	7.7	7.6	7.8
8.6	8.6	8.6	8.6
	8.9	9.7	
		10.4	
11.3	11.21	11.5	11.3
12.2	12.14	12.3	
13.3	13.3	13.8	13.5

The correspondence between the astronomical data and the model in Figure 6.15(b) can be interpreted as strong evidence for biology at redshifts $z \approx 0.83$, that is up to a distance $D \approx cz/H \sim 2.5\,\text{Gpc}$, assuming a Hubble constant of 100 km/s per Mpc. The new observations are consistent with the spread of microbial life encompassing a significant fraction of the radius of the observable Universe.

Over the past 20 years the strongest astronomical evidence for the biological model of interstellar dust within our Galaxy has been the data on the Galactic Centre source GC-IRS7. Since 1982 many attempts have been made to match the spectrum of this source in the 2–4 μm waveband using abiotically generated mixtures of organic materials. Irradiation of suitably constructed mixtures of inorganic ices has been shown to result in organic residues possessing spectra that 'fitted' the astronomical spectra to varying degrees (Tielens *et al.* 1996). The fits produced so far have left much to be desired, and moreover, all the arguments and comparisons presented thus far have begged the important question as to how the precise conditions under which the laboratory experiments were conducted could be reproduced with an unerring precision on a Galaxy-wide scale. To the end of his life, Fred Hoyle remained convinced, as indeed am I,

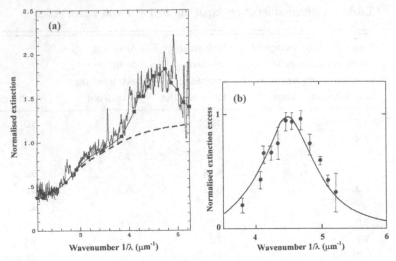

Figure 6.15 (a) The continuous line is the extinction curve for the gravitational lens
galaxy SBS0909+532 excluding well defined spectral lines due to MgII,
CIII and CIV, redshift $z = 0.83$ (Motta *et al.* 2002). The filled squares
joined by straight lines represent a least-squares fit to the data,
excluding well-defined spectral lines. The dashed curve is the scattering
background attributed to bacterial particles. (b) The curve is the
normalized absorption coefficient of an ensemble of 115 biological
aromatic molecules. The points are the observations of the galaxy
SBS0909+532 due to Motta *et al.* (2002).

that to seek alternative explanations of these data is futile, and reflects a deeply
held prejudice to shy away from an unpalatable conclusion: namely that life
had already evolved to an advanced stage long before its appearance on our
minuscule planet.

6.9   Entry into panspermia: a historical context

Fred Hoyle and I entered the panspermia arena, albeit with some trep-
idation, not through biology but via the astronomical arguments discussed in
earlier sections. From the mid 1980s onwards our progress in this direction
seemed inevitable and unstoppable, much to the chagrin of friends and colleagues. At the time the accepted theory of the origin of life was that due to
Haldane (1929) and Oparin (1953) in which it was proposed that life began in
a 'primordial soup' of organic chemicals that developed on a primitive Earth.
That theory had gained ground after the classic experiments of Miller and Urey
(1959) where it was shown that organic chemicals that may serve as the building
blocks of life could form from suitable mixtures of inorganic gases such as H_2O,

CH_4, NH_3 when subjected to electrical discharges or energetic radiation. But as we shall show in a later section these results are largely irrelevant to the origin of life itself.

Ideas concerning the existence of life outside the Earth have a great antiquity. They span many centuries and many different cultures. In most ancient philosophies of the Orient, for instance in Vedic and Buddhist writings, the cosmic character of life is taken for granted. It is regarded as an inherent property of the Universe that is itself infinite, timeless and eternal. A similar concept made only a brief appearance in Western Philosophy. The idea of living seeds or 'spermata' being ever present in the Cosmos was posited by the pre-Socratic philosopher Anaxagaras as early as the fifth century BC, but soon came to be replaced by an Aristotelian Earth-centred world view.

Until the late nineteenth century *panspermia* meant the passage of organisms through the Earth's atmosphere, not an incidence from outside the Earth. In this form it seems to have been used first by Lazzaro Spallanzani (1729–99). But almost a century before that, Francesco Redi had carried out what can be seen as a classic experiment in the subject. He had shown that maggots appear in decaying meat only when the meat is exposed to air, inferring that whatever it was that gave rise to the maggots must have travelled to the meat through the air.

A long wait until the 1860s then followed, until Louis Pasteur (1860) showed by experiments on the souring of milk and the fermentation of wine that similar results occurred when the agency passing through the air was bacteria, replicating as bacteria, not producing a visible organism like maggots. The world then permitted Pasteur to get away with a huge generalization, and honoured him greatly both at the time and in history for it: by then the world was anxious to be done with the old Aristotelian concept of life emerging from the mixing of warm earth and morning dew.

Pasteur's far-ranging generalization was to the precept that each generation of every plant or animal is preceded by a generation of the same plant or animal. This view was taken up enthusiastically by others, particularly by physicists among whom John Tyndall lectured frequently on the London scene, as for instance in a Friday evening discourse at the Royal Institution on 21 January 1870. It was to this lecture that the editorial columns of the newly established *Nature* (e.g. issue of 27 January 1870) objected with some passion. Behind the objection was the realization that were Pasteur's paradigm taken to be strictly true, the origin of life would need to be external to the Earth: for if life had no spontaneous origin, it would be possible to follow any animal generation by generation back to a time before the Earth existed, the origin being therefore required to be outside the Earth.

This was put in remarkably clear terms by the German physicist Hermann von Helmholtz (1872):

> It appears to me to be a fully correct scientific procedure, if all our attempts fail to cause the production of organisms from non-living matter, to raise the question whether life has ever arisen, whether it is not just as old as matter itself, and whether seeds have not been carried from one planet to another and have developed everywhere where they have fallen on fertile soil ...

The next facet in the story is associated with the Swedish chemist and Nobel laureate Svante Arrhenius, whose book *Worlds in the Making* appeared in English in 1908 (Arrhenius 1903, 1908). Arrhenius' contribution rested on two main points, one good, one not so good. The good point was that microorganisms possess unearthly properties, properties that cannot be explained by natural selection against a terrestrial environment. The example for which Arrhenius himself was responsible was the taking of seeds down to temperatures close to zero kelvin, and of then demonstrating their viability when reheated with sufficient care. Many other 'unworldly' properties have come to light over the years to which we shall have occasion to refer below.

The not-so-good point was that Arrhenius conceived of microorganisms travelling individually and unprotected through the Galaxy from star system to star system. He noticed that organisms with critical dimensions of 1 μm or less are related in their sizes to the typical radiation wavelengths from dwarf stars in such a way that radiation (light) pressure can have the effect of dispersing these particles throughout the Galaxy. But space-travelling individual bacteria would be susceptible to deactivation and damage from the ultraviolet light of stars, and this was already known in the first decades of the century. P. Becquerel (1924) mounted an attack on Arrhenius' views in 1924, on the basis of possible ultraviolet damage, and this attack was widely accepted and has been repeated many times since. But several other facts of relevance to this problem were not known at the time.

6.10   Extreme hardiness of bacteria

On the whole, microbiological research of the past 20 years has shown that bacteria and other microorganisms are remarkably space-hardy, far more than Arrhenius may have ever imagined (Postgate 1994). Microorganisms known as thermophiles and hyperthermophiles are present at temperatures above boiling point in oceanic thermal vents. Entire ecologies of microorganisms are present in the frozen wastes of the Antarctic ices. A formidable total mass of microbes exists in the depths of the Earth's crust, some 8 kilometres below the surface, greater than the biomass at the surface (Gold 1992). A species of a phototropic sulphur bacterium that can perform photosynthesis at exceedingly low light levels, approaching near total darkness, has been recovered from the Black Sea (Overmann *et al.* 1992). There are bacteria (e.g. *Deinococcus radiodurans*) that thrive in the cores of nuclear reactors. Such bacteria perform the amazing feat of using an enzyme system to repair DNA damage, in cases where it is estimated that the DNA experienced as many as a million breaks in its helical structure.

There is scarcely any set of conditions prevailing on Earth, no matter how extreme, that is incapable of harbouring some type of microbial life. As for ultraviolet damage under space conditions, this is very easily shielded against. A carbonaceous coating of only a few micrometres thick provides essentially a total shielding against ultraviolet light, and there are several modern experiments that have demonstrated precisely that. Next, let us note that many types of microorganisms are not really killed by ultraviolet light, they are only deactivated. It happens through a shifting of certain chemical bonds contained in the genetic structures of the organisms, without destroying the genetic arrangements themselves. This permits the original properties to be recovered once the ultraviolet radiation has been shut off. Furthermore, we know that microorganisms that are normally sensitive to ultraviolet light can, through repeated exposures, be made just as insensitive as the more resistant kinds – yet another unearthly property. Many other tests of the space hardiness of bacteria and viruses have recently been made. In one such test the bacterium *Bacillus subtlis* was exposed for nearly six years in space aboard NASA's Long Duration Exposure Facility and was found to retain viability. In other experiments currently in progress bacteria impacting sand with speeds of 0.3–0.8 km/s have been shown to survive (Rotan *et al.* 1998). These speeds are in comfortable excess of the terminal velocities of micrometre-sized particles following atmospheric braking. Experiments such as these show unequivocally that the transfer of microbes from a comet to Earth can take place without significant loss of viability.

6.11   Coalification of bacteria and interstellar organic molecules

Notwithstanding the remarks of the previous section, bacteria that have no protective coatings and which are exposed remorselessly to cosmic rays and to the background of starlight in open regions of interstellar space, in the so-called diffuse clouds, must be subject to degradation and eventual destruction. The process would be analogous to coalification of living material. Microorganisms expelled from any galactic source into unshielded regions of interstellar space will firstly become deactivated. Then the deactivated particles will be subject to steadily increasing degradation, ending in a release of free organic molecules and polymers, similar to what astronomers have been discovering since the late 1960s. The ultimate end product will be a transformation of a viable bacterium to a submicrometre-sized particle of coal, similar indeed to the material we have discussed in earlier sections.

Today an impressive array of interstellar molecules has been detected, and among them are a host of hydrocarbons: polyaromatic hydrocarbons of the type we have referred to earlier, the amino acid glycine, vinegar and the sugar glycolaldehyde (Hollis *et al.* 2000). Such organic molecules that pervade the interstellar clouds make up a considerable fraction of the available galactic carbon. Theories of how interstellar organic molecules might form via non-biological processes are still in their infancy, and in terms of explaining the available facts they leave much to be desired.

The overwhelming bulk of organic matter on the Earth is indisputably derived from biology, much of it being degradation products of biology. Might not the same processes operate in the case of interstellar organic molecules? The polyaromatic hydrocarbons that are so abundant in the Cosmos, as pointed out in Section 6.8, could have a similar origin to the organic pollutants that choke us in our major cities – products of degradation of biology, biologically generated fossil fuels in the urban case, cosmic microbiology in the interstellar clouds. The theory of cosmic panspermia that we have proposed leads us to argue that interstellar space could be a graveyard of cosmic life as well as its cradle. Only the minutest fraction (less than one part in a trillion) of the interstellar bacteria needs to retain viability, in dense shielded cloudlets of space, for panspermia to hold sway. Common sense dictates that this survival rate is unavoidable.

6.12   Replication properties of bacteria

By far the simplest way to produce such a vast quantity of small organic particles (with properties ranging from pristine bacteria to coals) and with sizes appropriate to bacteria is from a bacterial template. The power of bacterial replication is immense. Given appropriate conditions for replication, a typical doubling time for bacteria would be two to three hours. With a continuing supply of nutrients, a single initial bacterium would generate some 2^{40} offspring in 4 days, yielding a culture with the size of a cube of sugar. Continuing for a further 4 days, the culture, now containing 2^{80} bacteria, would have the size of a village pond. Another 4 days and the resulting 2^{120} would have the scale of a large comet. Yet another 4 days and the resulting 2^{160} bacteria would be comparable in mass to a molecular cloud like the Orion Nebula. And 4 days more still, a total of just 20 days since the beginning, and the bacterial mass would be that of a million galaxies. No abiotic process remotely matches this replication power of a biological template. Once the immense quantity of organic material in the interstellar material is appreciated, a biological origin for it becomes an almost inevitable conclusion.

6.13   Cometary panspermia

The sources of biological particles in interstellar clouds are comets, according to the Hoyle–Wickramasinghe theory. An individual comet is a rather insubstantial object. But our Solar System possesses so many of them, perhaps more than a hundred billion of them, that in total mass they equal the combined masses of the outer planets Uranus and Neptune, about 10^{29} grams. If all the dwarf stars in our Galaxy are similarly endowed with comets, then the total mass of all the comets in our Galaxy, with its 10^{11} dwarf stars, turns out to be some 10^{40} grams, which is just the amount of all the interstellar organic particles that is present in the dust clouds within the Galaxy.

How would microorganisms be generated within comets, and then how could they get out of comets? We know as a matter of fact that comets do eject organic particles, typically at a rate of a million or more tons a day. This was what Comet Halley was observed to do on 30–31 March 1986. And Comet Halley went on doing just that, expelling organic particles in great bursts, for almost as long as it remained within observational range. The particles that were ejected in March 1986 were well placed to be observed in some detail. No direct tests for a biological connection had been planned, but infrared observations pointed unexpectedly in this direction. The infrared emission spectrum of dust from Comet Halley obtained by Dayal Wickramasinghe and David Allen (1986) in March 1986 matched precisely the laboratory spectrum of bacterial grains as shown in Figure 6.16 (Wickramasinghe *et al.* 1986).

An independent analysis of dust impacting on mass spectrometers aboard the spacecraft Giotto also led to a complex organic composition, a composition that was fully consistent with the biological hypothesis. Broadly similar conclusions have been shown to be valid for other comets as well, in particular Comet Hyakutake and Comet Hale–Bopp. Thus one could conclude from the astronomical data that cometary particles, just like the interstellar particles, are *spectroscopically* identical to bacteria, existing in various combinations along with their degradation products.

The logical scheme for the operation of cometary panspermia is summarized in Figure 6.17. The dust in interstellar clouds must always contain the minutest fraction of bacteria (less than a trillionth) that retains viability despite the harsh radiation environment of space. This exceedingly modest requirement of survival would be utterly impossible to violate under any circumstances, so panspermia becomes inevitable. When a new star system (e.g. a solar system) forms from

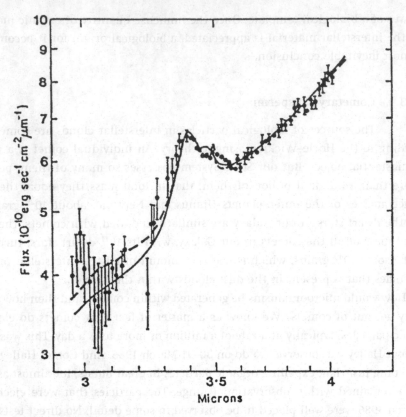

Figure 6.16 Emission by dust coma of Comet Halley observed by D. Wickramasinghe and D. A. Allen on 31 March 1986 (points) compared with bacterial models.

interstellar matter, comets condense in the cooler outer periphery as a prelude to planet formation. Each such comet incorporates at the very least a few billion viable bacterial cells, and these are quickly reactivated and begin to replicate in the warm interior regions of the comets, thus producing vast numbers of progeny. As a fully fledged stellar or planetary system develops, comets that plunge from time to time into the inner regions of the system would release vast quantities of bacteria in the manner discussed earlier for our own Solar System. Some of the evaporated bacterial material is returned into the interstellar medium. New stars and star systems form, and the whole cycle continues with a positive feedback of biologically processed material.

6.14   Improbability of life's origins: cosmic evolution

According to the ideas developed by Fred Hoyle and the author since 1980, life is considered to be a cosmic phenomenon: viable bacteria of cosmic

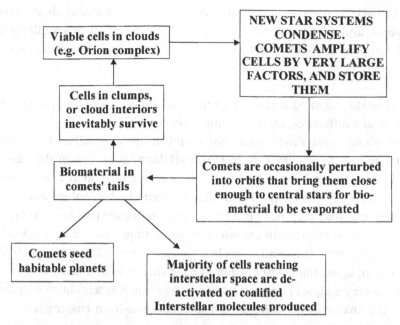

Figure 6.17 Cosmic amplification cycle of biology.

origin were transported to Earth to give rise to the range of terrestrial life that we find here. They were present already in the material from which the Solar System condensed, and their number was then topped up substantially by replication in cometary material. Thus the impacts of cometary material would have brought them to the Earth. The interiors of large-enough missiles are known to remain cool and relatively undisturbed by such impacts. The wiping out of resident cultures was then of no overall consequence because the cultures destroyed were replaced by new arrivals.

What now of the chemical processes in a warm little pond? Would they be capable of arriving at the molecular arrangements of such observed biological structures as DNA and RNA, or at the enzymes for which such structures code? A typical enzyme is a chain with about 300 links, each link being an amino acid of which there are 20 different types used in biology. Detailed work on a number of particular enzymes has shown that about a third of the links must have an explicit amino acid from the 20 possibilities, while the remaining 200 links can have any amino acid taken from a subset of about four possibilities from the bag of 20. This means that with a supply of all the amino acids supposedly given, the probability of a random linking of 300 of them yielding a particular enzyme is as little as

$$\frac{1}{(20)^{100} . 4^{200}} \approx 10^{-250}.$$

The bacteria present on the Earth in its early days required about 2000 such enzymes, and the chance that a random shuffling of already-available amino acids happens to combine so as to yield all the required 2000 enzymes is

$$2000![10^{-250}]^{2000},$$

which works out at odds of one part in about $10^{500\,000}$, with the factorial hardly making any difference, large as it might seem.

Everybody must surely agree that a probability as small as this cannot be contemplated. To a believer in the Oparin–Haldane paradigm of the warm little pond there has to be a mistake in the argument. Although it is known that the bacteria present on the Earth, almost from the beginning, were ordinary bacteria, everyday bacteria as one might say, it is argued that the first organisms managed to be viable with considerably fewer than 2000 enzymes (Mushegian and Koonin 1996). The number has been reduced from 2000 to 256. Additionally, one can imagine the lengths required of chains of amino acids to be reduced. Suppose for example one reduces the length as much as tenfold, to only 30 links. Then the chance of obtaining such a severely sawn-down enzyme is

$$256![10^{-25}]^{276}.$$

Neglecting the effect of the factorial, this amounts only to one part in 10^{6900}, still a probability that is wholly negligible even in the widest cosmic context. For comparison, there are about 10^{79} atoms in the whole visible Universe, in all the galaxies visible in the largest telescopes. This comparison shows in our opinion that life *must* be a cosmological phenomenon, not at all something which originated in a warm little terrestrial pond.

The best condition of all is for a spatially infinite Universe, a Universe that ranges far beyond the largest telescopes. Then the very small chance of obtaining a replicative primitive cell will bear fruit somewhere and, when it does, replication will cause an enormous number of the first cells to be produced, as I showed in the example of cometary interiors in Section 6.12. It is here that the immense replicative power of biology shows to great advantage, particularly since we can distribute the products of such replication over millions of galaxies. Each minute innovative step in the development of life – every gene – can generate and disperse enough copies of itself over a cosmic scale for a second highly improbable event to occur somewhere in one of the profusion of offspring. And so, by an extension of the argument, to the third, fourth, fifth improbable events: indeed, to a whole chain of improbable occurrences, which result at last in the magnificent range and variety of genes we have today, the genes that were already present at the formation of the Earth.

With the genetic components of life distributed widely throughout the Universe, it is a matter for each local environment to pick out arrangements that best fit the particular circumstances. In a case like the Earth, a complicated fitting together of the components occurred over the last several hundred million years, by a process that we recognize as evolution (Hoyle and Wickramasinghe 1981a,b).

On this model of the origin of life there would be little variation in the forms to which the process gives rise, at least so far as basic genes are concerned, over the whole of our Galaxy. Or indeed over all nearby galaxies. The rest of the story concerns the many ways in which the same basic genes can combine to produce rich varieties of living forms from one environment to another, always remembering that because of the large numbers involved – large numbers of stars, large numbers of planets and large numbers of galaxies – the system can afford many failures; for instance, the Earth would not have produced anything very noteworthy but for the events of the last half-billion years.

6.15   Decisive tests for panspermia

With a welter of evidence now available in support of a cosmic theory of life one might wonder why a direct experimental verification has not thus far been made. Since comets are the local carriers and amplifiers of microbial life, a direct examination of cometary material should be considered a project of high scientific priority.

NASA's STARDUST mission to recover material from comet P/Wild 2 was launched in February 1999. Its main task is to intercept the comet's tail in January 2004, trap comet dust onto aerogel plates and return them to Earth for analysis in January 2006. The mission was planned in the late 1980s at a time when hostility to panspermia was possibly at its height. In the event, no life-science experiments were planned, and the aerogel collection strategy involved high-speed collisions so that no cometary organics, let alone microbes, would be collected intact.

Less than one year into the STARDUST mission Franz Krueger and Jochem Kissel (2000) reported a serendipitous discovery. A mass spectrometer aboard STARDUST intercepted five interstellar dust particles. The shattered residue of these particles, impacting the instruments at a speed of 30 km/s, was found to include structures described as 'cross-linked hetero-aromatic polymers'. Such molecular fragments are consistent with the break-up products of bacterial cell walls, which are arguably the only types of structures that could survive the high impact speeds (Wickramasinghe et al. 2001).

Costly space programmes may not, however, be necessary to test the life-from-space hypothesis. Research centres in India, particularly the Tata Institute in Mumbai, have had a long track record in using balloon-borne instruments for astronomy – for example, cosmic-ray studies and infrared astronomy. For many years Fred Hoyle had tried to persuade the Indian scientists to use sterile balloon-borne instruments to retrieve cometary material in the high stratosphere. This was not done until January 2001, which was evidently the first time when the techniques were sufficiently well in place for absolutely aseptic collections to be guaranteed (Narlikar *et al.* 1998).

A number of specially manufactured aseptic stainless steel cylinders were evacuated to almost zero pressures and fitted with valves that could be opened and shut on ground telecommand. An assembly of such cylinders was suspended in a liquid neon environment to keep them at cryogenic temperatures, and the entire payload was launched into the stratosphere from the TATA Institute balloon launching facility in Hyderabad, India, on 20 January 2001. As the valves of the cylinders were opened at predetermined heights, ambient air rushed in to fill the vacuum and built up very high pressures within the cylinders. The valves were shut after a prescribed length of time, the cylinders hermetically sealed and parachuted back to the ground.

Back on the ground the cylinders were carefully opened and the air collected was made to flow through sterile membrane filters in a contaminant-free environment. Any bacteria or clumps of bacteria present in the stratosphere would then be collected on these filters. The laboratory analysis was conducted by a team of microbiologists in Cardiff, and initial results provided unambiguous evidence for the presence of clumps of viable cells in air samples from as high as 41 km (Harris *et al.* 2002). The detection was made using a fluorescent dye known as cyanine, which is taken up only by the membranes of living cells. When an isolate treated with the dye was examined under an epifluorescence microscope, pictures such as Figure 6.18 were obtained.

Similar pictures of cells have since been obtained after treating samples with a dye that is taken up only by DNA. So the presence of living cells replete with DNA is now confirmed in air samples collected at heights of 30–41 km. Particles such as are seen in this figure do not stay stationary at these heights. They fall under gravity at speeds of centimetres per second. From analysis of our samples we can conclude that the particles are falling through the stratosphere and arriving at the Earth's surface at the rate of about a ton per day over the entire planet. Because 41 km is well above the tropopause (about 16 km in the tropics), a level that acts as a virtual ceiling for aerosols lofted from the surface, the *prima facie* evidence is for an incidence of living bacterial cells from space. The manner in which the density of such particles varies with height also tends to

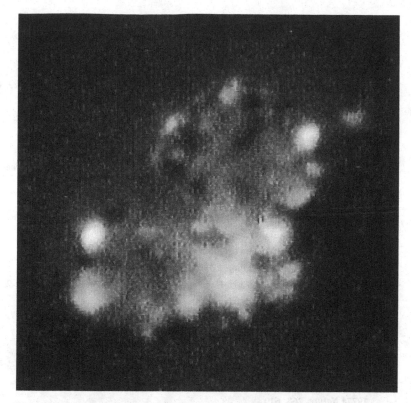

Figure 6.18 Clump of cells collected from the stratosphere, fluorescing in carbocyanine dye.

support the panspermia hypothesis, rather than the remote possibility that we have somehow sampled rare particles that were lofted from the ground. Whilst most of the recovered biomaterial appears to comprise cells that were deemed viable but not culturable, two cultures – one micrococcus and one microfungus – were recently obtained by Wainwright (Wainwright *et al.* 2002). Experiments of this kind are clearly of great relevance to establishing, once and for all, the credence of the Hoyle–Wickramasinghe panspermia hypothesis. Were the results we have obtained so far to be confirmed and validated in further experiments that are planned, one of the most controversial theories with which Fred Hoyle was involved would be vindicated.

References

ALLEN, D. A. & WICKRAMASINGHE, D. T. 1981 *Nature*, **294**, 239

AL-MUFTI, S. 1984 PhD Thesis, University College Cardiff

ARRHENIUS, S. 1903 *Lehrbuch de Kosmischen Physik*, Leipzig; *Worlds in the Making*, transl. H. Born, 1908

BECQUEREL, P. 1924 *Bull. Soc. Astron.*, **38**, 393

CATALDO, F., KEHEYAN, Y. & HEYMANN, D. 2002 *Int. J. Astrobiol.*, **1**, 79

CROVISIER, J. *et al.* 1997 *Science*, **275**, 1904

DANIELSON, R. E., WOOLF, N. J. & GAUSTAD, J. E. 1965 *ApJ*, **141**, 116

DONN, B. D., WICKRAMASINGHE, N. C., HUDSON, J. P. & STECHER, T. P. 1968 *ApJ*, **153**, 451

DULEY, W. W. & WILLIAMS, D. A. 1979 *Nature*, **277**

FORREST, W. J., GILLETT, F. C. & STEIN, W. A. 1975 *ApJ*, **192**, 351

FORREST, W. J., HOUCK, J. R. & REED, R. A. 1976 *ApJ*, **208**, L133

GEZARI, D. Y., SCHMITZ, M., PITTS, P. S. & MEAD, J. M. 1993 *Catalog of Infrared Observations* (NASA Reference Publ. 1294)

GOLD, T. 1992 *Proc. Nat. Acad. Sci.*, **89**, 6045

GUILLOIS, O. *et al.* 1999 *Solid Interstellar Matter: The ISO Revolution*, Les Houches, No. 11, (Les Ulis: EDP Sciences)

HALDANE, J. B. S. 1929 *The Origin of Life* (London: Chatto and Windus)

HARRIS, M. J. *et al.* 2002 *Proc. SPIE* **4495**, 192

HELMHOLTZ, H. VON 1872 in *Handbuch der Theoretischen Physik*, Vol. 1, Part 2, W. Thomson & P. G. Tait eds.

HOLLIS, J. M, LOVAS, F. J. & JEWELL, P. R., 2000 *ApJ*, **540**, L81

HOYLE, F. & WICKRAMASINGHE, N. C. 1962 *MNRAS*, **124**, 417

  1968 *Nature*, **217**, 415

  1970 *Nature*, **226**, 62

  1976 *Nature*, **264**, 45

  1977a *Nature*, **268**, 610

  1977b *Nature*, **270**, 323

  1979 *Ap&SS*, **66**, 77

  1980a *Ap&SS*, **68**, 499

  1980b *Ap&SS*, **69**, 511

  1980c *Ap&SS*, **72**, 183

  1981a *Evolution from Space* (London: J.M. Dent)

  1981b *Proofs that Life is Cosmic* (Institute of Fundamental Studies, Sri Lanka, Mem. 1)

  1984 *From Grains to Bacteria* (Cardiff: University College Cardiff Press)

  1988 *Ap&SS*, **147**, 245

  1989 *Ap&SS*, **154**, 143

  1991 *The Theory of Cosmic Grains* (Dordrecht: Kluwer Academic Publishers)

  1996 *Ap&SS*, **235**, 343

HOYLE, F., WICKRAMASINGHE, N. C., AL-MUFTI, S. & OLAVESEN, A. H. 1982a *Ap&SS*, **81**, 489

HOYLE, F., WICKRAMASINGHE, N. C., AL-MUFTI, S., OLAVESEN, A. H., & WICKRAMASINGHE, D. T. 1982b *Ap&SS*, **83**, 405

HOYLE, F., WICKRAMASINGHE, N. C. & AL-MUFTI, S. 1982c *Ap&SS*, **86**, 63

HULST, VAN DE, H. C. 1946–49 *Rech. Astron. Obs. Utrecht*, XI, Parts I and II

JOHNSON, F. M. 1971 *Ann. NY Acad. Sci.*, **194**, 3

JOHNSON, F. M. 1972 *Ann. NY Acad. Sci.*, **187**, 186

KNACKE, R. F., CUDABACK, D. D. & GAUSTAD, J. E. 1969 *ApJ*, **158**, 151

KRUEGER, F. R. & KISSEL, J. 2000 *Stern und Weltraum*, **5**, 330

MERRILL, K. M., RUSSELL, R. W. & SOIFER, B. T. 1976 *ApJ*, **207**, 763

MILLER, S. L. & UREY, H. C. 1959 *Science*, **130**, 245

MOTTA, V., MEDIAVILLA, E., MUNOZ, J. A., FALCO, E., KOCHANEK, C. S., ARRIBAS, S., GARCIA-LORENZO, B., OSCOZ, A. & SERRA-RICART, M. 2002 *ApJ* **574**, 719

MUSHEGIAN, A. S. & KOONIN, E. V. 1996 *Proc. Nat. Acad. Sci. USA*, **93**, 10268

NANDY, K. 1964 *Publ. Roy. Obs. Edinburgh*, **3**, 142

NARLIKAR, J. V. *et al.* 1998 *Proc. SPIE* **3441**, 301

OKUDA, H. *et al.* 1989 *IAU Symposium No 136*, 281
  1990 *ApJ*, **351**, 89

OORT, J. H. & VAN DE HULST, H. C. 1946 *BAN*, No. 376

OPARIN, A. I. 1953 *The Origin of Life*, transl. S. Margulis (Dover)

OVERMANN, J., CYOIONKA, H. & PFENNIG, N. 1992 *Limnol. Oceanogr.*, **33** (1), 150

PASTEUR, L. 1860 *C. R. Acad. Sci.*, **20**, 303

PENDLETON, Y. J. *et al.* 1994 *ApJ*, **437**, 683

POSTGATE, J. 1994 *The Outer Reaches of Life* (Cambridge: Cambridge University Press)

ROTAN, C. A. H. *et al.* 1998, *Proceedings of the Founding Convention of the Mars Society*, II eds. R. M. Zubrin and M. Zubrin

SAPAR, A. & KUUSIK, I. 1978 *Publ. Tartu Astr. Obs.*, **46**, 717

SCHALEN, C. 1939 *Uppsala Obs. Ann.*, **1**, No 2

STECHER, T. P. 1965 *ApJ*, **142**, 1683

STECHER, T. P. & DONN, B. D. 1965 *ApJ*, **142**, 1681

TAFT, E. A. & PHILLIPP, H. R. 1965 *Phys. Rev.*, **138A**, 197

TIELENS, A. G. G. M. *et al.* 1996 *ApJ*, **461**, 210

VANYSEK, V. & WICKRAMASINGHE, N. C. 1975 *Ap&SS*, **33**, L19

WAELKENS, C. & WATERS, L. B. F. M. 1997 in *From Stardust to Planetesimals*, eds. Y. J. Pendleton and A. G. G. M. Tielens, PASP Conference Series, p. 67

WAINWRIGHT, M. *et al.* 2002 *Proc. SPIE*, **4495**, 192

WHITTET, D. C. B. 1979 *Nature*, **281**, 708

WICKRAMASINGHE, N. C. 1967 *Interstellar Grains* (London: Chapman & Hall)
  1974a *Nature*, **252**, 462
  1974b *MNRAS*, **170**, 11p

WICKRAMASINGHE, N. C. & GUILLAUME, C. 1965 *Nature*, **207**, 366

WICKRAMASINGHE, N. C. & HOYLE, F. 1997 *Internet Journal Natural Science*, May
  1999 *Ap&SS*, **268**, 379

WICKRAMASINGHE, N. C., HOYLE, F. & AL-JABORI, T. 1989 *Ap&SS*, **158**, 135
  1990 *Ap&SS*, **166**, 333

WICKRAMASINGHE, N. C., WICKRAMASINGHE, A. N. & HOYLE, F. 1992 *Ap&SS*, **196**, 167

WICKRAMASINGHE, N. C., WICKRAMASINGHE, D. T. & HOYLE, F. 2001 *Ap&SS*, **275**, 181

WICKRAMASINGHE, N.C., LLOYD, D. & WICKRAMASINGHE, J.T. 2002a
   Proc. SPIE **4495**, 255
WICKRAMASINGHE, N.C., NARLIKAR, J.V., WICKRAMASINGHE, J.T. &
   WAINWRIGHT, M. 2002b *Proc. SPIE*, **4859**, 154
WICKRAMASINGHE, D.T. & ALLEN, D.A. 1986 *Nature*, **323**, 44
WICKRAMASINGHE, D.T., HOYLE, F., WICKRAMASINGHE, N.C. &
   AL-MUFTI, S. 1986 *Earth, Moon and Planets*, **36**, 295
WOOLF, N.J. & NEY, E.P. 1969 *ApJL*, **155**, L181

7

Worlds without end or beginning

JOHN D. BARROW

Centre for Mathematical Sciences, University of Cambridge

Fred Hoyle's greatest work was in the areas of nuclear astrophysics, galaxies and stellar evolution. But stop people on the street and ask them what they know him for and you will find that it is always for being a cosmologist, and a very particular sort of cosmologist at that: the co-inventor of something called the Steady-State universe. Although that work typecast him for the rest of his life, and often seemed to steer the direction of his work on all manner of non-cosmological subjects, it was only part of the cosmology he did. Ironically, reminiscent of John Wallis who managed to write the codebooks for both sides in the English Civil War, Hoyle also did outstanding work in founding the Big-Bang cosmological theory. Here, we want to take some snapshots of his contributions to cosmology with an emphasis upon ideas that still play an important role in cosmology, or which highlight ongoing discussions. Although the original Steady-State theory is no longer a viable cosmological theory, it contains ingredients which have re-emerged, transmogrified, in the most recent editions of the inflationary universe theory.

One of Einstein's greatest simplifying assumptions was to introduce the assumption that space can be assumed to be homogeneous and isotropic. This later became known more grandly as the 'Cosmological Principle'. Unfortunately, there is a surprising degree of confusion in the expository literature as to what this Principle actually is, and the nature of the primary observational evidence for it. In some textbooks it is presented as saying that the density of the universe is spatially uniform and isotropic. As a result, the discovery of Great Walls of galaxies or large-scale streaming motions has

The Scientific Legacy of Fred Hoyle, ed. D. Gough.
Published by Cambridge University Press. © Cambridge University Press 2004.

occasionally led to the idea that the Cosmological Principle might not be true after all. Others have taken seriously the idea that the density distribution in the universe might not be a homogeneous statistical process with a well defined finite mean. Instead, it is suggested that it might be a fractal. The answer to these puzzles is straightforward. The Cosmological Principle is just the statement that the metric of spacetime is well described by the homogeneous and isotropic metric first studied by Friedmann, Robertson, and Walker. At the Newtonian level the spacetime metric is just the gravitational potential Φ, and so the Cosmological Principle requires that the gravitational-potential perturbations $\delta\Phi/\Phi$ are small (Barrow 1989). This is not the same thing as requiring the density perturbations $\delta\rho/\rho$ to be small. For a density irregularity of size L with respect to the Hubble radius cH^{-1} the two are simply related by

$$\frac{\delta\Phi}{\Phi} \sim \frac{\delta\rho}{\rho}\left(\frac{L}{cH^{-1}}\right)^2. \tag{7.1}$$

Thus we see why it is that enormous cosmological density fluctuations on the scale of a single human being, $\delta\rho/\rho \sim 10^{30}$, cast no doubt on the validity of the Cosmological Principle: the associated gravitational potential perturbation is very small. The best direct evidence for the Cosmological Principle is supplied by the temperature isotropy of the microwave background radiation. The large-scale temperature anisotropy, $\delta T/T$, measures the gravitational-potential perturbations directly:

$$\frac{\delta\Phi}{\Phi} \sim \frac{\delta T}{T} \sim 2 \times 10^{-5}.$$

Thus it is the microwave background isotropy that supplies the principal evidence for the validity of the Cosmological Principle with our particle horizon.

In 1948 Hoyle, Bondi and Gold (HBG) first posed the question of what the universe might be like if the homogeneity of *space* assumed by the Cosmological Principle were extended to *spacetime*. This extension was named the Perfect Cosmological Principle. We now know that there are four possible 'universes' that are homogeneous spacetimes (their number could be increased if we were to allow oscillating closed universes but they would need to be the same on average in every cycle and this is unlikely given the effects of the second law of thermodynamics (Tolman 1931), or the cosmological constant, which inevitably brings the oscillations to an end (Barrow and Dabrowskii 1995)). There is the flat spacetime of Minkowskii – rather dull for a cosmological model. There is the Einstein static universe – first conceived by Einstein, with the help of the cosmological constant, to prevent solutions of general relativity from being *expanding* universes; but later he recanted and abandoned all interest in the cosmological constant owing to the instability of the static universe. There is Gödel's universe – rotating with the added attraction of time travel, but looking little like our universe. And last there is de Sitter's universe. This was the spacetime that HBG chose as the basis for a universe obeying the Perfect Cosmological Principle. The result became known as the Steady-State universe.

The spacetime of the Steady-State universe is described by the simple de Sitter metric

$$ds^2 = dt^2 - \exp[2H_0 t][dx^2 + dy^2 + dz^2]. \tag{7.2}$$

If we assume that this metric is a solution of Einstein's equations then the material content of the universe has constant density ρ given in terms of the constant Hubble expansion rate H_0 by the Friedmann equation

$$3H_0^2 = 8\pi G \rho = \text{constant}.$$

Notice that the two observables, the density and the expansion rate, are both epoch-independent: the Steady-State universe will look the same at all times as well as in all places, as the Perfect Cosmological Principle requires. There is no initial singularity where the density goes to infinity, and no observation should be able to distinguish any cosmic epoch from any other. This is therefore a very 'strong' theory, in Einstein's sense, with no room to manoeuvre in the face of adverse observational evidence. The only free parameter to be fixed by observation is H_0.

In the Steady-State theory (Hoyle 1948 and 1949) this universe was given an additional interpretation. It was argued that it required there to be a continuous creation of matter throughout spacetime in contrast to the sudden 'creation' of matter at $t = 0$ in the Big-Bang theory, as Hoyle first coined it, at a special time in the past. This idea apparently needed a modification of general relativity to introduce into Einstein's equations: a 'creation field' that created matter (from nothing) at exactly the rate needed to keep the density constant in the face of the expansion. This rate is very low and completely unobservable – about one atom per cubic metre in every 10 billion years. There is no chance of seeing it happen directly.

This idea was very stimulating and maintained a vigorous debate in theoretical cosmology for nearly 20 years, and led to the development of important observational tests of the Perfect Cosmological Principle by radio astronomers. However, I believe that there was no need for any continuous creation in the Steady-State universe at all. The only matter that 'appeared' in this universe was material coming through the horizon of observers. If we assume that matter is described by a perfect fluid with pressure p and density ρ then the de Sitter metric and Einstein's equations require that

$$\dot{\rho} + 3H(\rho + p) = 0.$$

Thus ρ is constant when the matter satisfies the equation of state

$$p_v = -\rho_v = \text{constant}.$$

This is equivalent to the presence of a cosmological constant or to a static scalar field with constant potential (McCrea 1951). Notice that this is the equation of state of a perfect fluid: there is no non-equilibrium behaviour, no broken world lines of 'created' particles, and everything is encompassed by the theory of general relativity with an appropriate stress tensor. The addition of other forms of matter (dust, radiation, magnetic fields) or spatial curvature has no significant effect upon this situation. These stresses just add rapidly falling density terms to the right-hand side of equation (7.2), and the exact de Sitter state is approached as time increases

$$3H^2 = 8\pi G(\rho_v + \rho_{\text{matter}}) \rightarrow 8\pi G \rho_v.$$

Seen in this light the whole acrimonious argument about the continuous creation of matter that was so problematic for the Steady-State theory appears to have been an example of much ado about nothing.

In retrospect, the Steady-State universe is an inflationary universe in which inflation always occurs. The most interesting aspects of the Steady-State investigations were into features of the Steady-State theory that we now recognise as attributes of an inflationary universe. In 1963 Hoyle and Narlikar gave the first example of a cosmic 'no hair' theorem when they proved that the de Sitter universe is stable against small inhomogeneous perturbations of the metric. Perturbations are all seen to die away by a geodesically moving observer. This allowed the Steady-State theory to explain why the universe was close to isotropy and homogeneity: the situation of very small metric potential perturbations needed to validate the Cosmological Principle was an inevitable consequence of the Steady-State metric. Hoyle and Narlikar (1963) conclude that:

> Providing the continuous creation of matter is allowed, the creation
> acts in such a way as to smooth out an initial anisotropy or
> inhomogeneity over any specified volume. Rotation in the sense of
> Gödel is never destroyed, but it is made arbitrarily small over the given
> proper volume. In other words, any finite portion of the universe
> gradually loses 'memory' of an initially imposed anisotropy or
> inhomogeneity...if the C-field is not present, the universe itself is

simply a 'transient' and the observed regularity is just 'chance'. If the
C-field is present...it seems that the universe attains the observed
regularity irrespective of initial boundary conditions.

By contrast, as they stress, the uniformity of the universe was something of mystery within the context of the standard Big-Bang theory. This mystery deepened after the discovery of the microwave background radiation, and was only dispelled after the idea of the inflationary universe emerged in 1981 (Guth). One can also see that the Steady-State theory predicted that the expansion of the universe should lie close to the critical state separating open from closed universes. Again, this was an unexplained coincidence within the standard Big-Bang theory. Unfortunately, these significant properties of the Steady-State theory never became a focus of the discussion, either by its adherents or its opponents.

The problems that emerged for the Steady-State theory were a mixture of theoretical and observational difficulties. The 'no hair' property of the de Sitter expansion becomes a liability if the expansion goes on forever as the Perfect Cosmological Principle dictates. All inhomogeneities get ironed out and there will be no stars and galaxies at all. One of the first challenges for Steady-State theorists was to find a mechanism for making galaxies and, although there were proposals to make them new in the gravitational wakes of other moving galaxies (Sciama 1959), no persuasive scenario emerged. Another difficulty, which Hoyle once told me was what worried him most about continuous creation, was the overt baryon asymmetry of the universe. Any form of continuous creation would naturally have produced equal amounts of matter and antimatter and a baryon-symmetric universe. Even today, when physicists do not believe that baryon number is conserved in Nature, it would be hard to obtain such a level of baryon non-conservation at the very low energies dictated by the Hubble scale H. Attempts by Hoyle and Narlikar (1966) to circumvent the baryon asymmetry led to a significant compromise of the Steady-State principle.

Another observational problem for a Steady-State universe is related to its infinite past. If there is a finite chance of anything happening it will happen infinitely often in a Steady-State universe. In particular, we might expect a Steady-State universe to team with living systems on our past light cone (Barrow and Tipler 1986). Or, as a graphic corollary, one might say that, if true, the Steady-State theory could not be original.

We have seen that the one free parameter in the Steady-State de Sitter metric that must be fixed by observation is H, the Hubble expansion rate of the universe. In the Steady-State theory this could, in principle, be anything. Moreover, since the age of the Steady-State universe is infinite, the quantity H^{-1} is not related to any finite 'age' of the universe. By contrast, in the Big-Bang theory, the quantity

H^{-1} is always of the same order as the expansion of the universe. Thus the fact that the observed value of H^{-1} is of order 10^{10} years, the main-sequence stellar lifetime is completely natural in a Big-Bang cosmology because we expect to be observing the universe soon after stable hydrogen-burning stars appear on the scene (Dicke 1957 and 1961). However, in the Steady-State theory it is a complete coincidence. There is no reason for there to be any link between the lifetime of stars and the expansion rate of the universe. This coincidence should have hinted to cosmologists that the Big-Bang theory was on the right track (Rees 1972). However, one must remember that one of the motivations for the Steady-State alternative was the apparent contradiction that once existed between the apparent age of the universe and the age of the Earth before the recalibration of the distance scale. Thus it would have been difficult to base a strong argument upon the proximity of H^{-1} to the stellar lifetime when it appeared at first that stellar lifetimes and planetary ages were actually much greater than H^{-1}.

In the end, the most persuasive observational evidence against the Steady-State cosmology was provided by observations of radio sources and the discovery of quasars, together with the discovery of the microwave background radiation. These observations, along with abundances of helium-4, showed that the universe was significantly different in the past: any Cosmological Principle was necessarily Imperfect. The universe is evolving from a hot dense environment into a cooler more rarefied one, and contains the essential fossil evidence predicted to be left over from the high-temperature beginnings by the Big-Bang theory. This evidence is generally interpreted to show that any steady state that exists in the universe must be non-local, and exist on a scale exceeding that of the particle horizon (about 20 billion light years) today.

Although Hoyle is usually associated with the development of the Steady-State theory, his most enduring contribution to cosmology is his work on the primordial nucleosynthesis of light elements. In 1964 Hoyle and Tayler wrote a key paper which showed how it was possible to predict the cosmological abundance of helium-4 and how the process was sensitive to particular aspects of cosmology and physics (like the number of light neutrino species). They concluded that either the universe had a hot singular origin or that the nucleosynthesis of helium in massive objects has been far more frequent than previously supposed. Subsequently, in 1967, Wagoner, Fowler and Hoyle subjected this problem to a fully detailed analysis that predicted all the abundances of elements up to oxygen, and they detailed the dependence of the abundances of deuterium, helium-3, helium-4 and lithium-7 on the entropy per baryon of the universe. This analysis has been fine-tuned, and the experimental nuclear and weak-interaction parameters have been updated as and when appropriate, but this is still essentially the standard 'textbook' exposition of the course of primordial nucleosynthesis that is a foundation stone of our understanding of the early universe.

During the past 15 years aspects of the Steady-State theory have re-emerged as components of the inflationary-universe scenario for the early stages of the universe. Inflation is just a finite period of cosmic history during which the expansion of the universe accelerates. The most likely behaviour is that in which the expansion of the universe is temporarily described by the de Sitter metric (7.2) of the Steady-State theory. This behaviour is precipitated by the expansion dynamics being temporarily dominated by the presence of a scalar field ϕ, with self-interaction potential $V(\phi) \geq 0$, so that its density and pressure are:

$$\rho = \frac{1}{2}\dot{\phi}^2 + V,$$

$$p = \frac{1}{2}\dot{\phi}^2 - V.$$

When the potential energy dominates the kinetic, then we have $p \approx -\rho \approx -V$, and Steady-State behaviour is well approximated. From this we can see why the density fluctuations produced on large scales from the inflation of the quantum fluctuations of the scalar field are expected to possess a particular 'constant curvature' form. If the universe is temporarily in a steady state, then it is a homogeneous spacetime and it must not be possible to distinguish the future from the past. In particular, it must not be possible to do this by using the amplitude of the metric potential fluctuations as a clock. Therefore they must all appear to an observer to have the same amplitude regardless of scale. By inspection of equation (7.1) we see that that is only the case when the density perturbations have a particular form:

$$\frac{\delta\rho}{\rho} \propto L^{-2}.$$

This is the famous Harrison–Zeldovich spectrum seen to high accuracy on large angular scales by the COBE satellite and from the ground. This argument shows why it arises in inflationary universes with a de Sitter phase.

It was Screaming Lord Sutch who first asked, 'Why is there only one Monopolies Commission?' Had he been a cosmologist he would probably also have asked why there is only one 'universe'. The inflationary universe theory has given rise to a cosmological perspective on which it is natural to expect our universe to be only locally isotropic and spatially homogeneous. Far beyond our visible horizon we now have a positive reason to expect that the geography of the universe is very irregular. Different small causally connected regions of the very early universe will have undergone different amounts of inflation and emerged with very different densities, fluctuation levels, and even numbers of fundamental forces and dimensions of space. More striking still, it has been argued by Vilenkin (1983) and Linde (1986) that inflation inevitably creates the conditions needed for further inflation to occur within sub-regions of inflated domains. This phenomenon of 'eternal' inflation appears to be never ending, although it is not yet known if it had a beginning in time. The global picture that emerges from eternal inflation is not dissimilar to Hoyle's later speculations about a global Steady-State universe permeated by local little bangs (Hoyle and Narlikar 1966). The eternal inflationary scenario predicts that most of the universe should still be inflating today and the whole is probably in a global steady state. We should expect to live in a local thermalized region in which inflation is no longer taking place. So far no one knows how to compute the probability distributions for outcomes in the eternal inflationary scenario and hence to estimate how 'likely' it is that anthropically hospitable domains arise that are large enough and cool enough for life to evolve within them.

This complication of our view of the history and geography of the universe extends well beyond the conditions of density and temperature we should find in space. We now understand how aspects of forces and constants of Nature might be expected to fall out differently from domain to domain of an eternally inflating universe. A full 'explanation' of why we find our domain to be as it is would always be left to say that some variables – perhaps the dimension of space, the value of the cosmological constant, or the number of forces of Nature – was the outcome of a random symmetry breaking process (Barrow 2002). If it had fallen out differently (for instance, if there had been more or fewer than three dimensions of space so that atoms could not exist) then 'observers' could not have existed. As with so many other things, Hoyle thought that way first, motivated by his discovery of the delicately positioned resonance levels in carbon-12 and oxygen-16 that allowed carbon and oxygen to exist simultaneously in the universe in anthropically useful abundances (Hoyle 1965):

> My inclination is to favour the view...that some, if not all, the dimensionless numbers in question are fluctuations; that is in other places in the universe their values would be different. The curious placing of the levels on ^{12}C and ^{16}O need no longer have the appearance of astonishing accidents. We can exist only in the portions of the universe where these levels happen to be correctly placed.

References

BARROW, J. D. 1989 *Quart. J. Roy. Astr. Soc.*, **30**, 163

    2002 *The Constants of Nature: From Alpha to Omega* (London: Jonathan Cape)

BARROW, J. D. & DABROWSKII, M. 1995 *MNRAS*, **275**, 850

BARROW, J. D. & TIPLER, F. J. 1986 *The Anthropic Cosmological Principle* (Oxford: Oxford University Press)

BONDI, H. & GOLD, T. 1948 *MNRAS*, **108**, 252

DICKE, R.H. 1957 *Rev. Mod. Phys.*, **29**, 355

  1961 *Nature*, **192**, 440

GUTH, A. 1981 *Phys. Rev. D*, **23**, 347

HOYLE, F. 1948 *MNRAS*, **108**, 372

  1949 *MNRAS*, **109**, 365

  1965 *Galaxies, Nuclei, and Quasars* (London: Heinemann)

HOYLE, F. & NARLIKAR, J. 1963 *Proc. Roy. Soc. A*, **273**, 1

  1966 *Proc. Roy. Soc. A*, **290**, 143

  1966 *Proc. Roy. Soc. A*, **290**, 162

HOYLE, F. & TAYLER, R.J. 1964 *Nature*, **203**, 1108

LINDE, A. 1986 *Phys. Lett. B*, **175**, 395

MCCREA, W.H. 1951 *Proc. Roy. Soc.*, **206**, 562

REES, M.J. 1972 *Comments Astrophys. Space Phys.*, **4**, 182

SCIAMA, D.W. 1959 *The Unity of the Universe* (London: Faber), Chapter 14

TOLMAN, R.C. 1931 *Phys. Rev.*, **37**, 1639

  1931 *Phys. Rev.*, **38**, 1758

VILENKIN, A. 1983 *Phys. Rev. D*, **27**, 2848

WAGONER, R., FOWLER, W. & HOYLE, F. 1967 *ApJ*, **148**, 3

8

Evolutionary cosmologies:
then and now

MALCOLM S. LONGAIR
Cavendish Laboratory, University of Cambridge

During the 1950s and early 1960s, the controversy between the proponents of Steady-State cosmology and those favouring evolutionary cosmologies overshadowed much cosmological discussion. Some aspects of the issues at stake are reviewed. While the debate was eventually decisively resolved in favour of the Big Bang cosmologies, the radio sources that were at the heart of the original controversy remain important cosmological probes. Some aspects of the most recent developments are described, particularly those related to the use of the radio galaxies in cosmology, and the role of the alignment effect in understanding how the relevant astrophysical processes have changed with cosmic epoch.

8.1   Personal memories

It was a great pleasure to be invited to talk at the meeting dedicated to the memory of Fred Hoyle. As readers may be aware, I come from the 'other side of the road', the Cavendish Laboratory, and, even worse, I was a student of Martin Ryle. As a result, my interactions with Fred had a somewhat different flavour from those of most of the other speakers. Nonetheless, I was on good terms with Fred.

I particularly remember the occasion of his 80th birthday, when I invited Fred to give a lecture to the Cavendish Physical Society as part of these celebrations. On 24 May 1995, Fred lectured to a packed audience on Steady-State Cosmology.

The Scientific Legacy of Fred Hoyle, ed. D. Gough.
Published by Cambridge University Press. © Cambridge University Press 2004.

He told me he was delighted to do so since he had made his first presentation of the theory in 1948 to the Cavendish Physical Society. I also remember his delightful remark that the only really big mistake he had made in his formulation of the theory was to call the creation field C, rather than ψ. If he had called it the latter, he would also have been remembered as the discoverer of cosmological inflation.

My own most powerful memory of Fred was of the postgraduate lecture course he gave in the Lent term of 1964, my first year as a research student in Cambridge. Fred entitled the course *Extragalactic Astrophysics and Cosmology* and it was given twice a week to a remarkable group of research students, many of whom went on to become leaders of astronomy. Fred would turn up with a few notes scribbled on what looked like the traditional envelope and run through what indeed turned out to be many of the key problems of astrophysics during the subsequent decades. Towards the end of the course, he tackled the problem of the origin of helium in the cosmos, reviewing the early work of Alpher, Gamow and Herman (Alpher and Herman 1948). Roger Tayler and John Faulkner were in the audience, and they realized that they could use the EDSAC-2 computer to carry out predictions of the cosmic helium abundance for a wide range of different models. In the course of the following two lectures, they unravelled in some detail the implications of these calculations, and the result was the famous *Nature* paper by Hoyle and Tayler (1964), which revived interest in the primordial synthesis of the light elements. This led subsequently to the standard results of these key cosmological calculations, carried out in collaboration with Fowler and Wagoner (Wagoner, Fowler and Hoyle 1967). It is not often that one has the luck to be present when a fundamental piece of astrophysical cosmology is developed in real time during a lecture course. It was a quite unforgettable experience and typical of Fred's unique approach to astrophysics.

My charge is to describe the controversy over the number counts of faint radio sources. This involved the proponents of Steady-State cosmology, Martin Ryle, who was head of the Radio Astronomy Group in the Cavendish Laboratory, and the Sydney radio astronomers led by Bernard Mills. Formally, my Ph.D. supervisor was Peter Scheuer, but, through the period 1963 to 1967, I was jointly supervised by Scheuer and Ryle; and indeed the subject of my dissertation concerned the joint themes that were closest to Ryle's heart astronomically, the astrophysical and cosmological evolution of powerful radio sources.

The facts of the case are now well documented in the literature. I particularly recommend the articles by Sullivan and Scheuer in the volume *Modern Cosmology in Retrospect* (Bertotti *et al.* 1990), which give fair descriptions of what actually happened, and the book by Sullivan *The Early Years of Radio Astronomy* (1984). I will summarize this history from my own perspective, and then tell the story of what happened subsequently, once the controversy had been decisively resolved in favour of the Big-Bang picture.

8.2   Cosmology in the immediate post-war era

The immediate post-war period was a turbulent era for cosmological theory. The standard models of general relativity had been discovered by Friedman in his famous papers of 1922 and 1924 and their content had been made known to the wider community through the advocacy of Lemaître in the 1930s, but the subject was nothing like the major industry it is today.[1] The important papers of Robertson and Walker in the 1930s elucidated the structure of the standard isotropic world models. The origin of the redshift was certainly not a settled issue and, in his papers on the redshift–magnitude relation, Hubble was cautious in interpreting the redshift as a consequence of the cosmological expansion. When I first studied cosmology, the standard work was Hermann Bondi's *Cosmology* (1960), which was a slightly amended version of the first edition published in 1952. As such, it provides a fascinating picture of the state of cosmological understanding through this turbulent epoch.

What is striking about Bondi's exposition is the wide range of cosmological theories that deserved serious attention. Besides the standard Friedman models, the Eddington–Lemaître models, Milne's Kinematic Cosmology, Dirac's theory involving large number coincidences and a variable gravitational constant, Eddington's fundamental theory and Steady-State theory are all there. In my reading of the book, there was a feeling that there was some fundamentally new physics to be uncovered through cosmological studies, but it was not at all clear what it would be.

The observations that provided the empirical basis for cosmological theories were in a relatively primitive state. There were only a few secure pieces of information about the large-scale features of the Universe as a whole. As Bondi emphasized, the fact that the sky is dark at night is a rather profound cosmological observation, showing that the Universe cannot be simultaneously isotropic, infinite, static and filled with stars. The second fact was Hubble's law, showing that the speeds of recession of galaxies are proportional to their distances. The third was evidence from the counts of faint galaxies that the Universe is isotropic and homogeneous on the large scale at about the 10% level. These observations left plenty of room for speculation about the large-scale dynamical behaviour of the Universe.

---

[1] I have given an account of the history of modern cosmology in my chapter 'Astrophysics and Cosmology' in *Twentieth Century Physics* (Longair 1995).

Lemaître (1931) had advocated a 'primaeval atom' realization of the earliest phases of the Friedman models, which he took to consist of a sea of closely packed neutrons. Among others, this idea was taken up by Gamow, who attempted to account for the synthesis of all the chemical elements by primordial nucleosynthesis during the early expansion of the Universe from this dense state. The motivations for this proposal were, first, the fact that the abundances of the chemical elements seemed to be remarkably uniform among the population of stars and, second, the inference from the theory of stellar structure that their central temperatures were not high enough to result in nucleosynthesis. The computations of Alpher and Hermann (1948) showed, however, that Gamow's programme would not work. The lack of a stable isotope with mass number eight presented a formidable barrier for the synthesis of the heavy elements in the early phases of the expansion.

Fred's remarkable discovery of the triple-alpha resonance (Hoyle 1953) and its experimental verification by Ward Whaling and his colleagues (Dunbar et al. 1953) played a key rôle in the story. Fred had unlocked the key to the understanding of the processes of nucleosynthesis in stars and soon he, the Burbidges and Fowler showed that the chemical elements could be synthesized in stars (Burbidge et al. 1957). Primordial synthesis of the elements therefore became irrelevant. The proponents of Steady-State cosmology asked the question: 'What observational evidence is there that the Universe passed through a hot, dense phase?' In the 1950s, the answer was clearly: 'None'.

The persuasive advocacy of Fred, Hermann Bondi and Thomas Gold caught the imagination of the professionals and the public alike. In the theory, the professionals saw a way of avoiding the timescale problem, which was besetting the standard Friedman picture (Bondi and Gold 1948, Hoyle 1949). All the standard models with zero cosmological constant have ages less than H_0^{-1}, and Hubble's value of the Hubble constant H_0 resulted in an age of the Universe less than about 2 billion years. This age was in decisive conflict with geological estimates of the age of the Earth, which were about 4.5×10^9 years. During the 1950s, the value of the Hubble constant was revised downwards by about a factor of two by Walter Baade, making it a much closer run thing, but still uncomfortably close.

Rereading Bondi's book, the evidence was quite evenly balanced between the Steady-State and the Big-Bang pictures. For many astrophysicists and cosmologists, the biggest problem with the Steady-State theory was the need for the continuous spontaneous creation of matter to replace the matter that is constantly driven apart by some unknown field – this was the price that had to be paid for eliminating the initial singularity of the standard Big Bang. Nonetheless, many professionals and amateurs were attracted by the concept of a Universe, infinite in space and time, which preserved the same appearance for all fundamental observers for all time, as embodied in the *Perfect Cosmological Principle* of Steady-State theory. It was into this hotbed of controversy that the extragalactic radio sources made an unexpected and dramatic entrance.

8.3   Radio astronomy and cosmology

Radio astronomy had its origins in Jansky's announcement of the discovery of radio emission from the plane of the Galaxy in May 1933. His discovery and its follow-up by Reber in the early 1940s had little impact upon the astronomical community. The development of radar during the Second World War had, however, two immediate consequences for radio astronomy. First, sources of radio interference that might confuse radar location had to be identified. In 1942, James S. Hey and his colleagues at the Army Operational Research Group in the UK discovered intense radio emission from the Sun, which coincided with a period of unusually high sunspot activity. Then, in 1946, immediately after the War, his group discovered the first discrete source of radio emission, which lay in the constellation of Cygnus, the source that became known as Cygnus A. The second consequence was that the extraordinary research efforts to design powerful radio transmitters and sensitive receivers for radar resulted in new technologies, which were to be exploited by the pioneers of radio astronomy, all of whom came from a background in radar.

After the War, a number of these scientists, including Martin Ryle, Bernard Lovell and Stuart Pawsey, began the systematic study of these radio astronomical phenomena that had been discovered more or less by accident. Further discrete sources of radio emission were discovered and radio interferometry provided the best means of measuring their positions with improved accuracy. In 1948, Martin Ryle and Graham Smith discovered the strongest radio source in the Northern Hemisphere, Cassiopeia A, and in 1949 the Australian radio astronomers John Bolton, Gordon Stanley and Bruce Slee succeeded in associating three of the radio sources with remarkable nearby astronomical objects, the supernova remnant known as the Crab Nebula and the strange nearby galaxies NGC 5128, associated with the source Centaurus A, and M87, associated with Virgo A. In addition to the diffuse radio emission of our own Galaxy, these early surveys established the existence of a population of discrete radio sources, some concentrated towards the plane of the Galaxy, but many others lying outside it. There was some uncertainty as to whether the isotropic component of the source population was associated with nearby radio stars in our own Galaxy or with distant extragalactic objects.

The radio astronomers could not answer this question from the radio observations alone since the radio data did not provide any distance measure for the sources. Distances could only be determined by finding an associated optical object and measuring its distance. In 1951, Graham Smith measured the positions of the two brightest sources in the northern sky, Cygnus A and Cassiopeia A, with an accuracy of about 1 arcmin, and this led to their optical identification by Walter Baade and Rudolph Minkowski using the Palomar 200-inch Telescope. Cassiopeia A was associated with a young supernova remnant in our own Galaxy, while Cygnus A was associated with a faint, distant galaxy at a redshift of 0.056. The latter observation immediately showed that the radio sources could be used in cosmological studies. Fainter radio sources would lie at significantly greater cosmological distances, and hence probe the Universe at epochs earlier than the present.

Initially, Martin Ryle supported the view that the discrete radio sources were 'radio stars' in our own Galaxy, but the identification of radio sources with galaxies such as M87 and NGC 5128, and especially the identification of Cygnus A, persuaded him to change his mind, and from then on he adopted what turned out the be the correct view that the vast majority of the radio sources observed in directions away from the Galactic Plane are distant extragalactic objects.

To find radio sources at cosmological distances, more sensitive radio surveys were needed, and this is where Ryle's genius as a physicist, radio scientist and engineer shone through. His deep understanding of the principles of radio interferometry led to the discovery of the principle of *aperture synthesis*, which enabled high angular resolution and sensitivity to be obtained in radio astronomical observations. For many years, Ryle and his colleagues ploughed a lone furrow in advocating the power of aperture synthesis as one of the most powerful means of advancing radio astronomy. The depth of his insight is well testified by the ubiquitous use of aperture synthesis as the route to high resolution and sensitivity in radio astronomy today, as exemplified by instruments such as the VLA and the VLBA.

To carry out a deeper survey of the radio sky, Ryle and Antony Hewish designed and constructed a large four-element interferometer operating at 81.5 MHz, which was sensitive to small angular diameter radio sources. The second Cambridge (2C) survey of radio sources was undertaken in 1954 and the first results published in the following year (Shakeshaft *et al.* 1955). Ryle and his colleagues found that the small diameter radio sources were uniformly distributed over the sky (Figure 8.1), and that the numbers of sources increased enormously as the survey extended to fainter and fainter flux densities (Figure 8.2). In any uniform Euclidean model, the numbers of sources brighter than a given limiting flux density S are expected to follow the relation $N(\geq S) \propto S^{-3/2}$, whereas, at the faintest flux densities, Ryle found an excess of faint sources which could be described by $N(\geq S) \propto S^{-3}$. He concluded that the only reasonable

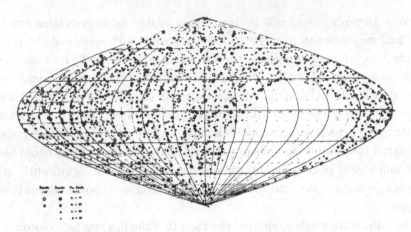

Figure 8.1 The distribution of small angular diameter radio sources from the 2C
survey in galactic coordinates (Shakeshaft *et al.* 1955). The blank area of
sky lies in the Southern Hemisphere, and was unobservable from
Cambridge.

interpretation of these data was that the sources were extragalactic, that they
were objects similar in luminosity to Cygnus A, and that there was a much
greater number density of sources at large distances than nearby, contrary to
the precepts of the Steady-State cosmology. As Ryle expressed it in his Halley
Lecture at Oxford in 1955 (Ryle 1955):

> This is a most remarkable and important result, but if we accept the
> conclusion that most of the radio stars are external to the Galaxy, and
> this conclusion seems hard to avoid, then there seems no way in which
> the observations can be explained in terms of a Steady-State theory.

These remarkable conclusions came as a surprise to the astronomical community. There was enthusiasm, but also scepticism that such profound conclusions
could be drawn from the counts of radio sources, particularly when their physical nature was not understood and only the brightest 20 or so objects had
been associated with relatively nearby galaxies. Just like the COBE observations
40 years later, these results hit the front page of the quality newspapers. The
deep impression made by the ensuing controversy is illustrated by the fact that,
after public lectures on cosmology today, I am regularly asked about the debate
between the Steady-State and Big-Bang theories.

The Sydney group were making radio surveys of the southern sky at about
the same time with the Mills Cross, and they found that they could represent
the number counts by the relation $N(\geq S) \propto S^{-1.65}$, which they argued was not
significantly different from the expectation of uniform world models. In 1957,
Bernard Mills and Bruce Slee stated:

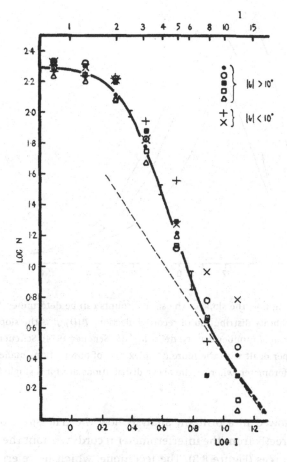

Figure 8.2 The integral number counts $N(\geq S)$ of radio sources in the 2C survey
(Shakeshaft *et al.* 1955). The dashed line shows the predictions of a
uniform Euclidean world model, $N(\geq S) \propto S^{-1.5}$. The observed counts are
very much steeper than this relation, showing a large excess of faint
radio sources.

We therefore conclude that discrepancies, in the main, reflect errors
in the Cambridge catalogue, and accordingly deductions of
cosmological interest derived from its analysis are without foundation.
An analysis of our results shows that there is no clear evidence for any
effect of cosmological importance in the source counts.

The problem with the Cambridge number counts was that they extended to
surface densities of sources such that the flux densities of the faintest sources
were overestimated because of the presence of even fainter sources in the beam
of the telescope, a phenomenon known as *confusion*, which was poorly understood at that time. The hero of this part of the story was undoubtedly Peter

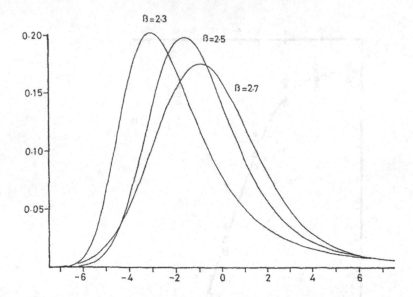

Figure 8.3 Illustrating how the slope of the source counts can be determined from the probability distribution of record deflections $P(D)$. β is the slope of the *differential* number counts: $dN \propto S^{-\beta} dS$ (Scheuer 1974). Scheuer's 1957 paper dealt with the more complex case of observations made with an interferometer, whereas the above distributions are for a single beam telescope.

Scheuer, who in 1957 showed in a brilliant analysis how the true slope of the counts could be found directly from the interferometer records without the need to identify individual sources (Figure 8.3). The technique, which he referred to as the $P(D)$ technique, showed that the slope of the source counts was actually -1.8. Ironically, this result, which we now know is exactly the correct answer, was not trusted, partly because the mathematical techniques used by Scheuer were somewhat forbidding. I also remember him telling me that nobody believed him, Ryle because he did not find $N \propto S^{-3.0}$, and Mills because he did not find $N \propto S^{-1.5}$. The dispute reached its climax at the Paris Symposium on Radio Astronomy in 1958, and the conflicting positions were not resolved (Bracewell 1959).

The resolution only came with further surveys, the 3C and 4C surveys, which were much less susceptible to the effects of source confusion and which enabled accurate positions and optical identifications with distant galaxies to be made. These optical identification programmes led to the discovery of quasars in the early 1960s. The combined 3C and 4C source counts showed an excess over the expectations of Euclidean world models (Gower 1966), showing that Ryle's conclusions of 1955 were correct, but that he had significantly overestimated the magnitude of the excess. Nonetheless, the effect was still a large one – the discrepancies with the Friedman and Steady-State models are very much greater than these simple comparisons suggested because the predicted source counts for a uniform world model converge very rapidly as soon as the source populations extend to significant cosmological redshifts (Scheuer 1975, Longair 1998). By the mid 1960s, with the large increase in the numbers of sources identified with distant galaxies and quasars in these bright samples, the evidence was compelling that there was indeed a large excess of sources at large redshifts, and this was at variance with the expectations of Steady-State cosmology.

By the early 1960s, relations between Fred and Ryle had reached a very low ebb. In the mid 1960s, Peter Scheuer and I made a serious effort to get Fred and Ryle together to see if there was common ground. By this time, Scheuer and I thought that the issues were rather clear cut and that we would all benefit from some form of reconciliation. The four of us met in the old lecture room in the Institute of Astronomy. We talked for about 45 minutes, but there was no meeting of minds – Fred and Ryle simply reiterated their somewhat entrenched views. Fred, Ryle and Scheuer are now all dead. It will remain with me as one of the saddest moments of my scientific career.

8.4   Evolutionary cosmologies today

A lot of water has passed under the bridge since those turbulent days of the 1950s and 1960s. The case for adopting Big-Bang models to describe the large-scale dynamics of the Universe is quite overwhelming. Let me summarize briefly some of the key pieces of evidence.

- As deep surveys of radio sources at different frequencies became available, the same features of the number counts found in the 3C and 4C surveys were observed at all wavelengths, as is illustrated by the compilation by Jasper Wall (Figure 8.4). All the number counts exhibit the same generic features, with a steep source count at high flux densities and strong convergence at low flux densities. In the very deepest counts, which extend to microjansky flux densities, a new population of nearby star-forming galaxies is observed.

- In order to determine the nature of the evolutionary changes with cosmic epoch, it is necessary to know the redshifts of the objects counted. This proved to be a highly non-trivial task since the galaxies associated with even the bright radio sources are faint and lie at cosmological distances. For example, 3C295 is among the brightest 10 extragalactic radio sources in the northern sky and has a redshift of

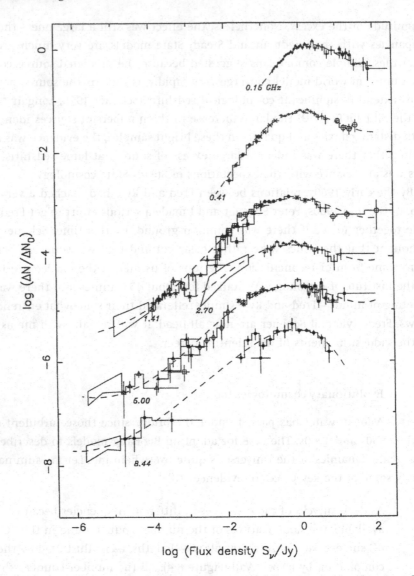

Figure 8.4 The normalized differential counts of radio sources at a wide range of different frequencies (Wall 1996). The number beside each number count is the frequency of the survey in GHz.

0.46 – for many years this was the most distance galaxy known. The determination of the evolution of the luminosity functions for radio sources has consequently been a long and arduous business, but the general features of satisfactory evolution functions that can account for the observed counts all involve very strong changes in the comoving space density of radio sources with cosmic epoch. A typical

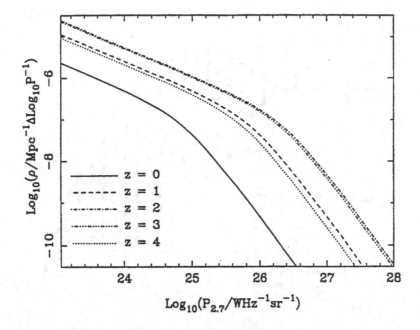

Figure 8.5 A realization of the change in the form of the luminosity function of strong radio sources by Dunlop (1994). In this luminosity-evolution model, the radio luminosity function is shifted to greater radio luminosities out to redshifts $z \sim 2$–3 and then begins to shift back.

realization of the evolution function as determined by James Dunlop and his colleagues is shown in Figure 8.5. The steep number count results in an enormous change in the comoving number density of sources of a given radio luminosity with cosmic epoch, or redshift, as can be seen in Figure 8.5.

- Further direct evidence for strong evolutionary effects with cosmic epoch is provided by the large quasar surveys carried out as part of the large 2dF project being carried out at the Anglo-Australian Telescope. The evolution of the optical luminosity function for over 6000 quasars shows the same sort of strong evolutionary effects observed in the radio source population, but these quasars are now wholly selected at optical wavelengths (Figure 8.6).

- The counts of faint galaxies show a large excess of faint blue galaxies relative to the expectations of the standard world models (Figure 8.7). The excess is less pronounced at the longer red and infrared wavelengths. The large number of faint blue galaxies at large redshifts is spectacularly demonstrated by the image of the Hubble Deep Field. Even a casual glance at that image shows that the faint population of

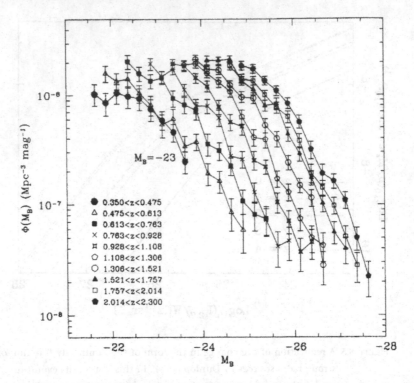

Figure 8.6 The evolution of the optical luminosity function of optically selected quasars over the redshift interval $0.35 \leq z \leq 2.3$ from the 2dF QSO Redshift Survey. Over 6000 quasars are included in these data (Boyle *et al.* 2002). This form of evolution would be consistent with a luminosity-evolution picture.

galaxies is very different from those observed in the nearby Universe. There are vastly more compact blue galaxies and also many more peculiar and distorted galaxies. It is natural to interpret these observations as star-forming galaxies within the context of hierarchical clustering models of galaxy formation.

- These observations can be used to work out how the rate of star formation has changed with cosmic epoch. All observational determinations show that the rate of star formation was about an order of magnitude greater at a redshift of one than it is at the present epoch. What took place at greater redshifts is still the subject of some debate since the optical data need substantial corrections to take account of the effect of dust obscuration. The submillimetre determinations of the change in the star formation rate with cosmic epoch are free from this problem and show that the star formation rate remained at a high level to much larger redshifts.

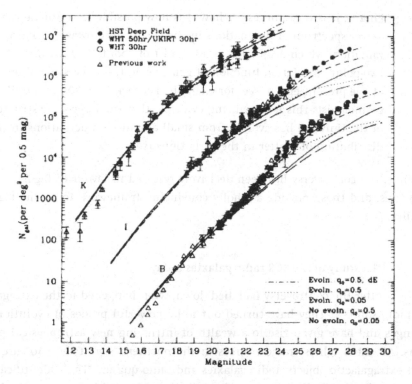

Figure 8.7 The counts of faint galaxies in the B, I and K wavebands compared with the expectations of uniform world models. The counts follow closely the predictions of uniform world models at magnitudes less than about 21, but there is an excess of galaxies in the B and I wavebands at fainter magnitudes (Metcalfe *et al.* 1996).

- Undoubtedly, however, the most powerful evidence that the Universe passed through a hot dense phase has been the isotropy and spectrum of the cosmic background radiation (Bennett *et al.* 1996, Fixsen *et al.* 1996). Not only does the remarkable isotropy of the radiation provide a secure observational validation for the Robertson–Walker models, but the thermal spectrum also demonstrates that the Universe must have passed through a hot, dense equilibrium phase.

- Equally compelling is the evidence that the light elements, ^{4}He, ^{3}He, D and ^{7}Li were synthesized in the early stages of expansion of the Universe. There is no explanation for these elements in terms of nucleosynthesis inside stars, but they are naturally created by primordial nucleosynthesis. Thus, Gamow's programme has in the end proved crucial for cosmology, but not quite in the way he had intended.

- Finally, it is very impressive how it is now possible to reconcile the power spectrum of fluctuations in the cosmic microwave background radiation, which originated at a redshift of about 1000, with the two-point correlation function of galaxies on the large scale observed at the present epoch (see, for example, Peacock *et al.* 2001). For all cosmologists this is convincing evidence that the large-scale structure of the Universe has evolved from small amplitude fluctuations in the distribution of matter in the early Universe.

Thus, the controversy has been decisively resolved in favour of Big-Bang cosmologies, and these provide a wholly convincing framework for cosmological studies.

8.5   The study of the 3CR radio galaxies today

After the controversy had died down, what happened to the extragalactic radio sources? They have turned out to be powerful probes of evolutionary changes and have given rise to a wealth of intriguing new astrophysical problems. All the brightest radio sources in the northern sky at $|b| \geq 10°$ are distant extragalactic objects, radio galaxies and radio quasars. Their identification proved to be a difficult task since the galaxies are very faint, but, by good fortune, the optical spectra of many of the very faint radio galaxies contain strong, narrow emission lines. Even more remarkably, the strength of the emission lines increases with radio luminosity and so, for a flux density-limited sample such as the 3CR sources, it was no more difficult to measure the spectra of the largest redshift galaxies than those of their low redshift counterparts. This enabled Spinrad and his colleagues to measure redshifts for most of the faint radio galaxies in the 3CR sample by the early 1980s.

It turned out that the radio quasars and radio galaxies span exactly the same range of redshifts, and their radio luminosity functions exhibit precisely the same form of evolution with cosmic epoch. One of the more remarkable aspects of the 3CR radio galaxies and radio quasars is that orientation-based unification schemes are successful in accounting for their properties. Every analysis we have made of their properties, such as their cosmological evolution and the statistics of their numbers, physical sizes and asymmetries, is consistent with the simplest orientation-based unification schemes, and so it is natural to assume that the host galaxies of radio quasars are galaxies like the radio galaxies.

In the early 1980s, Simon Lilly and I made infrared K-magnitude observations of a complete sample of the 3CR radio galaxies and found the intriguing result that they have a remarkably well defined K-magnitude–redshift relation (Lilly and

Longair 1984). The galaxies at redshifts $z \sim 1$ were about a magnitude brighter than expected if the stellar populations of the galaxies had remained unchanged with cosmic epoch. If, however, account is taken of the passive evolution of their stellar populations, it is found that the galaxies should have been about a magnitude brighter at these redshifts for world models with $\Omega_0 = 1$, $\Omega_\Lambda = 0$. This seemed to suggest that we would be able to obtain information both about the evolution of the optical, infrared and radio properties of the radio source population, as well as information about cosmological parameters from these studies. We carried out surveys of substantial samples of sources at a wide range of frequencies and flux densities, and this culminated in the paper by Dunlop and Peacock (1990).

A major spanner was thrown in the works, however, when Chambers *et al.* (1987) and McCarthy *et al.* (1987) discovered that the optical images of the faint 3CR radio galaxies were aligned with their radio axes, the effect known as the *alignment effect*. This indicated that the radio source activity was influencing the optical and, possibly, infrared images, and all bets were off until we had a better understanding of the nature of the effect.

Fortunately, having been an interdisciplinary scientist on the Science Working Group for the Hubble Space Telescope, I was guaranteed 44 hours of HST observing time, and I devoted all of it to observations of the 28 brightest radio galaxies in the northern sky in the redshift interval $0.6 \leq z \leq 1.8$, with a view to understanding the origin of the strong cosmological evolutionary effects and the importance of the alignment effect. Philip Best, Huub Röttgering and I were able to assemble complete sets of optical HST images, infrared images at 2 μm from the UK Infrared Telescope and radio images from the VLA for all the sources in the sample. The results were quite remarkable.

They are best illustrated by considering the eight radio galaxies in the redshift interval $1 \leq z \leq 1.3$. These all have roughly the same intrinsic radio luminosities and so they can be compared quite independently of the cosmological model. Figure 8.8 shows that they all display the strong alignment effect. In the figure, the radio galaxies are shown in order of increasing separation of their radio source components. In the standard picture of the evolution of double radio sources, they are powered by jets from the active nucleus and, for sources of the same luminosity, the greater the separation of the double source components, the greater the age of the source. In the left-hand column of Figure 8.8, the HST observations are shown with angular resolution about 0.1 arcsec, while in the right-hand column, the UKIRT K-images are shown with angular resolution about 1 arcsec. The infrared images show the old stellar populations of the radio galaxies, which are all giant elliptical galaxies. In contrast, the HST images are dominated by the aligned emission, stimulated by the passage of the radio jet,

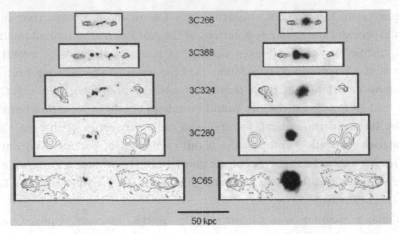

Figure 8.8 Illustrating the alignment effect (Best *et al.* 1998). These five radio
galaxies are shown on the same physical scale. The images on the left
are from the HST and the right-hand images are in the K-waveband as
observed with the UKIRT. On both diagrams, the radio contours have
been superimposed. The sources are shown in order of increasing radio
size. These are the smallest of the eight 3CR radio galaxies in the sample
in the range $1 \le z \le 1.3$.

and quite overwhelming the stellar populations of the galaxies. It can be seen
that the alignment effect is strongest in the most compact sources, with optical
knots aligned along the radio source axis. As the radio source becomes larger
and larger, the emission regions are confined to the galaxy itself rather than
extending into the surrounding intergalactic medium.

At this point, a great deal of detailed analysis had to be carried out to understand how the optical and infrared images could be influenced by the emission
associated with the alignment effect, and whether or not there was any contam-
ination of the infrared magnitudes by compact nuclear components. It turned
out that there were only two cases in which there was evidence that the infrared
magnitudes were being influenced by a nuclear component. In no case was there
a significant contribution of the aligned emission to the K-magnitudes. After a
great deal of analysis of the data, a similar K–z relation for 3CR radio galaxies
to that which Simon Lilly and I had derived was found (Figure 8.9). This was,
however, far from the end of the story, because Eales and Rawlings (1996) had
derived a K–z relation for radio galaxies from their 6C radio survey which was
about six times fainter than the 3CR sample. It turned out that, at redshifts
$z \ge 1$, the 6C galaxies are about 0.8 mag fainter than the 3CR galaxies. Their
K–z relation suggested that the 6C galaxies exhibited little or no increase in
K-luminosity at early epochs.

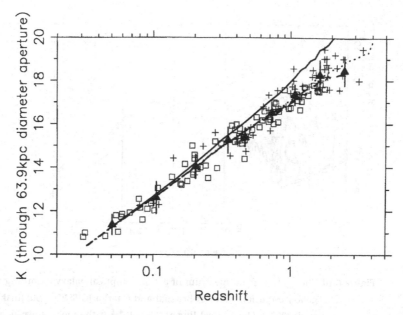

Figure 8.9 The *K–z* relation for powerful 3CR (□) and 6C (+) radio galaxies. (Inskip *et al.* 2002a).

There seemed no question, however, but that we were observing evolution of the stellar populations of the galaxies in the 3CR sample. A better test of the evolution of the stellar populations was carried out by comparing the location of the radio galaxies on the fundamental plane for elliptical galaxies with giant elliptical galaxies at the present day. This involved plotting the surface brightness of the galaxies against their de Vaucouleurs radii. We found that, once the effects of passive evolution of their stellar populations were taken into account, the 3CR radio galaxies lay precisely along the fundamental plane for giant elliptical galaxies.

8.6   New data, new analyses

We have returned to these problems with a matched set of observations of 6C radio galaxies.

8.6.1   *The K–z relation revisited*

First of all, we have carried out a new analysis of the *K–z* relation for a much wider range of world models (Inskip *et al.* 2002a). This was stimulated by the fact that the preferred cosmological model nowadays is not the $\Omega_0 - 1$, $\Omega_\Lambda = 0$ model, but rather models with finite cosmological constant. The other major improvement was the use of the latest galaxy evolution codes of Bruzual

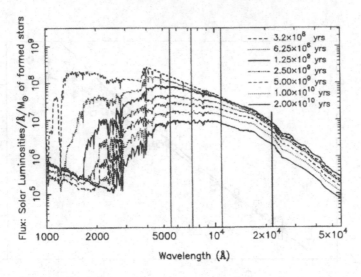

Figure 8.10 The ageing of the spectrum of a giant elliptical galaxy according to the galaxy evolution codes of Bruzual and Charlot (GISSEL 2000) (Inskip *et al.* 2002a). The vertical line at 2.2×10^4 Å is the mean K-wavelength. The vertical lines to the left of this line show the rest wavelengths of the emission of galaxies observed in the K-waveband at redshifts $z = 1, 2$ and 3.

and Charlot (GISSEL 2000). Using these codes, we were able to test in much more detail than before a very much wider range of possible changes of the properties of the radio galaxies with cosmic epoch. For example, we were able to study changes in the epoch of star formation, the metallicity of the galaxies, and the possibility that starbursts associated with the radio source events might influence the K-luminosities of the galaxies.

The new Bruzual and Charlot galaxy evolution codes enable us to understand a number of intriguing features of the evolution of galaxies and how they affect the predictions of different cosmological models. The important diagram is Figure 8.10, in which the passive evolution of the ultraviolet–optical–infrared spectrum of a standard giant elliptical galaxy is shown at different ages. It is assumed that all the stars are formed at some large redshift. The right-hand vertical line shows the 2.2 μm K-waveband and the vertical lines to the left of it show the locations of the rest wavelengths, which are observed in the K-waveband at redshifts of 1, 2 and 3. To the left of the left-most vertical line, it can be seen that the spectrum changes dramatically, while the behaviour of the galaxy spectrum to the right of the lines is remarkably systematic. Indeed, the evolution at a particular wavelength can be well described by simple analytical functions. This behaviour at wavelengths longer than about 500 nm reflects the fact that the predicted spectral evolution is well determined by the evolution of stars with mass

$M \leq 2 M_\odot$. The predominant factor in determining this spectral evolution is the evolution of main-sequence stars onto the giant branch, and this depends primarily upon the slope of the initial mass function over this rather narrow mass range. In fact, very good estimates of the K-corrections for elliptical galaxies of different ages can be made by reading the spectral luminosities from Figure 8.10 for the preferred ages of the world models at different redshifts. The stability of these predictions means that 2.2 μm is a very good archaeological waveband for understanding the global evolution of stellar populations out to $z = 3$.

In our study of the $K-z$ relation, we have shown that, because of the rather stable behaviour of the integrated spectrum of giant elliptical galaxies, the choice of cosmological model is more important than the stellar evolution corrections, which came as something of a surprise.

In our previous work, we had shown that the observed $K-z$ relation could be naturally explained in cosmological models with $\Omega_0 = 1$, $\Omega_\Lambda = 0$, once account had been taken of the effects of stellar evolution, as illustrated by the dashed lines in Figure 8.9. In the new analysis, we have considered a wide range of other models, including those with a finite value of Ω_Λ, in particular, the favoured model with $\Omega_0 = 0.3$, $\Omega_\Lambda = 0.7$. The results of this analysis are best appreciated from Figure 8.11, which shows the predicted relations relative to

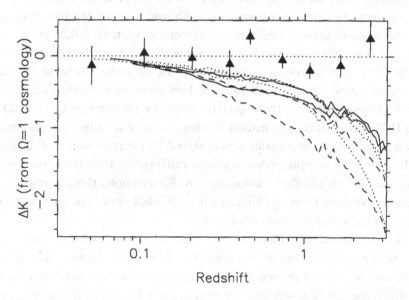

Figure 8.11 Differences in the $K-z$ relation for giant elliptical galaxies for a range of different cosmological models, normalized to the model with $\Omega_0 = 1$, $\Omega_\Lambda = 0$ (horizontal dotted line) (Inskip et al. 2002a). The models with $\Omega_0 = 0.3$, $\Omega_\Lambda = 0.3$ and $\Omega_0 = 0.3$, $\Omega_\Lambda = 0.7$ are indicated by solid lines, the later model having the more negative values of ΔK. All models include passive evolution of the stellar populations of the galaxies.

those of the $\Omega_0 = 1, \Omega_\Lambda = 0$ model, which is shown by the horizontal dotted line passing though the observed mean relation. The lower solid line shows the predicted K–z relation for the preferred model with $\Omega_0 = 0.3, \Omega_\Lambda = 0.7$, and it is clearly a poor fit to the data. The simplest way of understanding these differences is that, in the $\Omega_0 = 1, \Omega_\Lambda = 0$ model, the effective distance of the galaxy is the smallest at a given redshift. In all the other models with $\Omega_0 < 1, \Omega_\Lambda > 0$, the distances of the galaxies are greater than in this model and so, at a given redshift, they must be intrinsically more luminous for a given K-magnitude. Thus, the literal interpretation of Figure 8.11 is that, at large redshifts, the radio galaxies were intrinsically more luminous than those at the present epoch.

This is a somewhat surprising result since the radio galaxies are associated with the most luminous galaxies and, at a redshift $z = 1$, are as luminous as the brightest galaxies in clusters. According to the preferred hierarchical clustering picture of galaxy formation, the galaxies at large redshifts should be *less* massive than those nearby. The observed increase in luminosity might be due to a number of other causes: for example, the galaxies might be more luminous because they had lower metallicity, or the redshift at which the bulk of the stellar populations were formed might have been only about $z = 3$, rather than at a very much larger redshift. In addition, the K–z relation might have been affected by star formation associated with the radio source event. We have investigated all these possibilities, particularly for the model with $\Omega_0 = 0.3, \ \Omega_\Lambda = 0.7$, but none of these effects seem capable of accounting for the discrepancies seen in Figure 8.11. It is interesting why starbursts are of limited value in minimizing the discrepancies. This arises because we have good limits on the ultraviolet star-forming luminosities of these galaxies from the photometry carried out with the HST, which strongly constrain the amount of star formation that could be taking place when the galaxies were observed. Our best guess is that the galaxies at large redshift are just more luminous intrinsically than those nearby. There are a number of possible reasons for this. For example, there is evidence that these galaxies are likely to be located in somewhat richer cluster environments than the corresponding galaxies at $z \ll 1$.

The astrophysical problem is similar to that of accounting for the fact that super-massive black holes with masses $M \sim 10^9 \, M_\odot$ must be present in quasars at very large redshifts. The same result is found in the luminous 3CR radio galaxies in which the rate of generation of energy in particles and magnetic fields is close to the Eddington limit for a $10^9 \, M_\odot$ black hole. According to the Press–Schechter formalism, the massive galaxies that host these super-massive black holes can barely have formed at the largest redshifts at which the powerful radio galaxies and quasars are observed.

8.6.2   Understanding the origin of the strong evolutionary effects

The problem with a flux-limited sample such as the 3CR sample is that it is impossible to separate out effects that are correlated with radio luminosity from those associated with cosmic epoch. This so-called degeneracy can be broken by carrying out observations of a matched sample of fainter radio sources that span the same redshift range as the 3CR radio galaxies. To achieve this, we have carried out a similar programme of observations for a matched sample of 6C radio sources, which are about six times fainter than the 3CR sources.

The most striking differences between the sources in the combined 3CR and 6C samples concern the dynamics of the emission line regions and the ionization state of the gas (Inskip *et al.* 2002b). Both are strongly correlated with the physical size of the double radio source. The differences are clearly observed in an ionization diagnostic diagram in which the line ratios of different species are plotted, which are sensitive to the excitation mechanism of the emission line regions (Figure 8.12). On the diagram, the areas occupied by regions that are

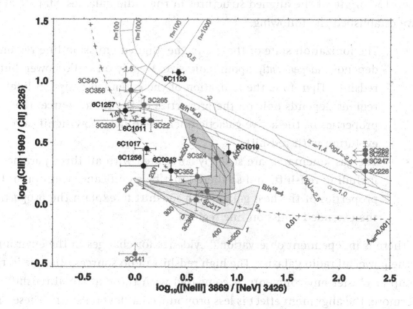

Figure 8.12 An ionization diagnostic diagram for the emission line regions associated with the alignment effect. The grey lines towards the top of the diagram show the expectations of photoionization models, while the shaded areas towards the lower middle of the diagram show the expectations of shock ionization models. The small double radio sources are predominantly in the latter region, whereas the large double sources tend to lie towards the region occupied by the photoionization models (Inskip *et al.* 2002b).

excited by photoionization and by shock excitation are indicated. The sense of the diagram is that the small radio sources (≤ 120 kpc) exist in a lower ionization state, consistent with the expectations of the shock ionization models of Dopita and Sutherland (1996), while the larger sources (≥ 120 kpc) are consistent with photoionization being the excitation mechanism. Shock excitation in the small sources is also consistent with their much greater velocity widths, as compared with the larger sources.

Such analyses provide insight into the environments of the radio galaxies at $z \sim 1$, and consequently into the cause of the strong evolutionary effects seen in the radio source population. In order to create the intense emission line regions observed in these sources, there must be a large population of cool clouds in the vicinity of the radio galaxy with a very small filling factor $\varepsilon \sim 10^{-6}$. The origin of these clouds is not clear, but they might result from cooling flows, or from interactions between galaxies. In either case, they provide evidence relevant to the fuelling of the black holes in the nuclei of these galaxies.

The new data on the 6C sample enable us to answer some of the questions about the origin of the aligned structures in the radio galaxies. Statistically, we have established the following.

- The ionization state of the gas in the aligned emission line regions depends *independently* upon: radio size, weakly on radio power, but not redshift. Therefore, the ionization of the extended emission line regions depends only on the properties of the radio source; the properties of the active galactic nucleus do not show significant evolution with redshift.

- The gas kinematics are strongly correlated with all three parameters, including redshift, and so there must be significant evolution of the properties of the host galaxy with redshift to explain the properties of the extended emission line regions.

There is independent observational evidence for changes in the environment of the powerful radio galaxies. The high redshift radio sources often lie in richer group or cluster environments as compared with those at small redshifts. Furthermore, the alignment effect is less pronounced at low redshifts. These results suggest that the distribution and density of gas clouds in the vicinity of the radio sources varies with redshift and that interactions between the surrounding interstellar and intergalactic medium and the radio source are less important at low redshifts. Furthermore, it is clear that the gas kinematics is influenced by the power of the radio source, specifically by the jet luminosity which powers the outer radio lobes. In turn, the expanding radio sources may alter the distribution of cool gas clouds in the intergalactic medium, and the extent of the observed aligned emission. The shocks observed in the aligned emission may also trigger star formation.

8.7   Concluding remarks

The study of the astrophysical and cosmological evolution of extragalactic radio sources has changed dramatically since its controversial early history. No one could have predicted that they would display such a remarkable set of properties that provide important information relevant to a wide range of topics in high energy astrophysics and astrophysical cosmology. As Hermann Bondi remarked following my lecture, he had always suspected that understanding the cosmological evolution of the extragalactic radio sources was to be a complex and lengthy story. He was absolutely correct – the bonus is that we have learned a great deal about a range of new astrophysical processes, which could scarcely have been dreamed of when this story began.

References

ALPHER, R. A. & HERMAN, R. C. 1948 Nature, 162, 774

BENNETT, C. L., BANDAY, A. J., GORSKI, K. M. et al. 1996 ApJ, 464, L1

BERTOTTI, B., BALBINOT, R., BERGIA, S. & MESSINA, A. (eds.) 1990 Modern Cosmology in Retrospect (Cambridge: Cambridge University Press)

BEST, P. N., LONGAIR, M. S. & RÖTTGERING, H. 1998 MNRAS, 295, 549

BONDI, H. 1952, 1960 Cosmology (Cambridge: Cambridge University Press)

BONDI, H. & GOLD, T. 1948 MNRAS, 108, 252

BOYLE, B. J., CROOM, S. M., SHANKS, T., OUTRAM, P. J., SMITH, R. J., MILLER, L. & LOARING, N. S. 2002 in A New Era in Cosmology, eds. N. Metcalfe and T. Shanks (Astron. Soc. Pacific Conf. Proceedings, San Francisco) 283, p. 72

BRACEWELL, R. N. (ed.) 1959 Paris Symposium on Radio Astronomy (Stanford: Stanford University Press)

BRUZUAL, G. & CHARLOT, S. 2000 GISSEL 2000 Spectral Synthesis Code

BURBIDGE, E. M., BURBIDGE, G. R., FOWLER, W. A. & HOYLE, F. 1957 Rev. Mod. Phys., 29, 547

CHAMBERS, K. C., MILEY, G. K. & van BREUGEL, W. J. M. 1987 Nature, 329, 624

DOPITA, M. A. & SUTHERLAND, R. S. 1996 ApJS, 102, 161

DUNBAR, D. N. F., PIXLEY, R. E., WENZEL, W. A. & WHALING, W. 1953 Phys. Rev., 92, 649

DUNLOP, J. S. 1994 in The Frontiers of Space and Ground-Based Astronomy, 27th ESLAB Symposium, eds. W. Wamsteker, M. S. Longair and Y. Kondo (Dordrecht: Kluwer Academic Publishers) p. 395

DUNLOP, J. S. & PEACOCK, J. A. 1990. *MNRAS*, **247**, 19

EALES, S. A., RAWLINGS, S. 1996 *ApJ*, **460**, 68

FIXSEN, D. J., CHENG, E. S., GALES, J. M., MATHER, J. C., SHAFER, R. A. & WRIGHT, E. L. 1996 *ApJ*, **473**, 576

GOWER, J. F. R. 1966 *MNRAS*, **133**, 151

HOYLE, F. 1949 *MNRAS*, **109**, 365

1953 *ApJS*, **1**, 121

HOYLE, F. & TAYLER, R. J. 1964 *Nature*, **203**, 1108

INSKIP, K. J., BEST, P. N., LONGAIR, M. S. & MACKAY, D. J. C. 2002a *MNRAS*, **329**, 277

INSKIP, K. J., BEST, P. N., RÖTTGERING, H. J. A., RAWLINGS, S., COTTER, G. & LONGAIR, M. S. 2002b *MNRAS*, **337**, 1407

LEMAÎTRE, G. 1931 *Nature*, **127**, 706

LILLY, S. M. & LONGAIR, M. S. 1984 *MNRAS*, **211**, 833

LONGAIR, M. S. 1998 *Galaxy Formation* (Berlin: Springer-Verlag)

1995 'Astrophysics and Cosmology' in *Twentieth Century Physics*, eds. L. M. Brown, A. Pais and A. B. Pippard, **3**, 1671 (Bristol: IOP Publishing and Philadelphia New York: AIP Press)

MCCARTHY, P. J., VAN BREUGEL, W. J. M., SPINRAD, H. & DJORGOVSKI, S. 1987 *ApJ*, **321**, L29

METCALFE, N., SHANKS, T., CAMPOS, A., FONG, R. & GARDNER, J. P. 1996 *Nature*, **383**, 236

MILLS, B. & SLEE, O. B. 1957 *Aust. J. Phys.*, **10**, 162

PEACOCK, J. A., COLE, S., NORBERG, P. *et al.* 2001 *Nature*, **410**, 169

RYLE, M. 1955 *Observatory*, **75**, 137

SCHEUER, P. A. G. 1957 *Proc. Camb. Phil. Soc.*, **53**, 764

1974 *MNRAS*, **167**, 329

1975 in *Stars and Stellar Systems, vol. 9 of Galaxies and the Universe*, eds. A. R. Sandage, M. Sandage and J. Kristian (Chicago: University of Chicago Press)

SHAKESHAFT, J. R., RYLE, M., BALDWIN, J. E., ELSMORE, B. & THOMSON, J. H., 1955 *Mem. RAS*, **67**, 106

SULLIVAN, W. T. III 1984 *The Early Years of Radio Astronomy* (Cambridge: Cambridge University Press)

WAGONER, R. V., FOWLER, W. A. & HOYLE, F. 1967 *ApJ*, **148**, 3

WALL, J. V. 1996 in *Extragalactic Radio Sources*, IAU Symposium No. 175, eds. R. Ekers, C. Fanti and L. Padrielli (Dordrecht: Kluwer Academic Publishers) p. 547

9

Alternative ideas in cosmology

JAYANT V. NARLIKAR

Inter-University Centre for Astronomy and Astrophysics, Pune

It is shown that in cosmology, as in the rest of science, the evolution of observational evidence and theoretical concepts has often led to the acceptance of ideas that were once considered outlandish. Fred Hoyle himself was responsible for generating several such ideas, although he was much ahead of his time. Here some of those ideas are outlined, ideas that were thought to be unrealistic at the time they were proposed, but which have now been assimilated into mainstream cosmology. A general comment that emerges from such examples is that highly creative individuals who are far ahead of their time do not get the recognition they deserve once their ideas are accepted as standard.

9.1   Introduction

The original title suggested for my contribution to the meeting dedicated to the memory of Fred Hoyle had been 'Alternative cosmologies'. I changed it to the present title for the following reason. There have been occasional reviews of alternative cosmologies from time to time, the most recent one being by myself and Padmanabhan (2001). Fred himself had been the originator of a few alternative cosmologies. However, in the present context, while highlighting his contributions to cosmology it is more interesting to look at ideas *on specific topics* of cosmological interest rather than at cosmological models per se.

In the decade following the observations of the microwave background radiation by Penzias and Wilson (1965), most cosmologists made up their minds that the so-called Big-Bang cosmology provides the correct framework in which

The Scientific Legacy of Fred Hoyle, ed. D. Gough.
Published by Cambridge University Press. © Cambridge University Press 2004.

to describe the history and large-scale structure of the Universe. Although there existed a range of possible models within this framework, they become collectively known as 'standard cosmology'. Any other cosmological model that did not fall within the standard framework was identified as 'non-standard cosmology'. A review entitled 'Non-standard cosmologies' by myself and Kembhavi (1979) summarized how such models dealt with the various theoretical concepts in cosmology and the available observational data.

However, in the 1980s the adjective 'non-standard' (which has somewhat negative connotations) was replaced by 'alternative', it being presumed that the alternative was with respect to standard cosmology.

Nevertheless, the adjective 'standard' is itself a misnomer, since it implies something stable and invariant with respect to which others can be measured. (The dictionary meaning of 'standard' is typically 'of recognized authority or prevalence'.) In a highly critical article on cosmology Disney (2000) has given reasons as to why this adjective is inappropriate in the sense in which it is used in the context of cosmology. Indeed, he has argued that the standard model in cosmology is nowhere as secure and laboratory-tested as the standard model in particle physics.

That the standard cosmology has evolved considerably in the four decades since the 1960s is seen from the various new concepts it has acquired during that period, namely, inflation, dark matter, non-baryonic matter, cosmological constant or quintessence or dark energy, new physics involving more than four dimensions, etc. Indeed, ideas that were once considered 'non-standard' have now become acceptable as 'standard'.

It makes sense therefore to concentrate here on such alternative ideas in cosmology, particularly those with whose origin Fred Hoyle was associated. As I propose to show, these ideas were not accepted at the time they were proposed, but gradually found their way into the 'standard' category as cosmology evolved. Before proceeding further, however, it is instructive to take a brief look at the historical evolution of cosmological ideas from earlier times.

9.2   From Aristarchus to Hubble

For viewing history one may define 'standard' at any epoch as representing ideas believed by the majority at that epoch, while 'alternative' would stand for ideas significantly different from the standard one (prevailing at the time) but believed by a minority. As will shortly be seen, this distinction between 'standard' and 'alternative' is epoch-dependent. An idea falling within the alternative basket at epoch t might well be transferred to the standard basket at a later epoch $t + \Delta (\Delta > 0)$. Naturally, if $\Delta \gg$ human life-span, then we have to conclude that the originator(s) of the alternative idea(s) failed to get credit for them while still alive.

Take Aristarchus (*c.* 310–230 BC), for example, who argued that the Earth not only spins about its axis, but it also goes around the Sun. To prove his point, he proposed what is today known as the *parallax method* for measuring distances of nearby stars. The test failed not because the underlying hypothesis was false but because the stellar distances were underestimated and the observational techniques were not sensitive enough to measure the actual parallax.

The standard cosmology of those times firmly believed a geocentric universe. That belief remained intact till Copernicus (AD 1473–1543) appeared on the scene. However, it took nearly a century before his alternative idea of the heliocentric cosmology became accepted. Thus for Copernicus, $\Delta \cong 100$ yr. So far as Aristarchus is concerned, his contribution has been belatedly recognized relatively recently. His bust in his home-town of Samos carries the inscription:

> Aristarchus of Samos...320–230 BC...First to discover the Earth
> revolves around the Sun...Copernicus copied Aristarchus 1530 AD...

Shall we say that Δ(Aristarchus) $\cong$ 23 centuries?

While the heliocentric theory became standard, it too had its factual limitations. By the beginning of the twentieth century, Herschel's picture of the Sun close to the centre of the Galaxy had become standard, and was adopted by J. C. Kapteyn in what had become known as the 'Kapteyn universe'. Harlow Shapley's measurements of distributions and distances of globular clusters, however, led to the realization that the Sun must be well away ($\sim$8–10 kpc) from the Galactic Centre. In this case one may set Δ at $\sim$10 years, and Shapley could get the credit during his lifetime.

An alternative idea that roused considerable passion and controversy began with Immanuel Kant (1724–1804), and is known as Kant's 'island universe' hypothesis. This hypothesis envisaged the Universe as a limitless system, like a vast ocean, in which galaxies like our Milky Way existed like islands. Observations had indeed revealed a number of nebulae, cloud-like faint but luminous systems, some of which (like the Andromeda Nebula) were claimed by a small minority to be distant galaxies, the Kantian islands. The standard view to which even Shapley subscribed was, however, that *everything* observed till that date was part of our Milky Way, which was envisaged as the Great Galaxy. The above alternative claim was dismissed as unfactual. See, for example, the following extract from the popular 1905 book called *The System of the Stars* by Agnes Clerke:

The question whether nebulae are external galaxies hardly any longer needs discussion. It has been answered by the progress of research. No competent thinker, with the whole of the available evidence before him, can now, it is safe to say, maintain any single nebula to be a star system of co-ordinate rank with the Milky Way...

The note of finality and certainty indicates the confidence felt by the majority in the correctness of the standard picture. Within two decades, however, this view had to be abandoned in the face of growing evidence against it and the lack of data in its favour. Let us first look at the latter.

The more accurate method of measuring the distances using the period–luminosity relation for the Cepheids, which had been introduced earlier, was applied by Hubble (1926) to M31 and M33. He found 50 variables in M31, of which 40 were cepheids, and 35 cepheids in M33. In addition, there were known to be nine in NGC6822, and Shapley had measured 105 in the Small Magellanic Cloud (SMC).

Of course, Hubble realized and stressed that these distances would be affected by any change in the zero point of the period–luminosity relation as it had been derived by Shapley. Still, the very large distances derived for the spirals were compatible with many attempts to determine the proper motion of the spirals, which had always led to null results (a velocity of 1000 km s^{-1} transverse to the line of sight would correspond to an annual proper motion of the order of $0''.0007$, a value far below what could be measured).

However, there was a stumbling block which led many to question these large distances. This arose from the claims by A. van Maanen (1916–30) at Mount Wilson to have determined proper motions *in the spiral arms* of a number of nearby spiral galaxies including M33, M81, M101, NGC2403, 4051, 4736, 5055, and 5195. At an indicative distance of 10^6 light years an annual motion $\sim 0''.01$, which was of the order of what was being claimed, corresponds to a velocity of $\sim 15\,000$ km s^{-1}. Thus van Maanen's results corresponded to periods of rotation of the spirals, if they were assumed to have sizes similar to the Milky Way, of the order 10^7 years or less, implying ejection of matter on similar timescales, so that the spirals would disintegrate in times of this order. However, later observations could not confirm these claims, and they gradually faded away.

Side by side with the lack of confirmation of proper motions, the new method of measuring distances using the Cepheid variables led Hubble to the conclusion that nebulae like the Andromeda Nebula (M31), M33, and several others lie well beyond the Galaxy. And so the Kantian island universe hypothesis began to gain credibility. If we date this change of paradigm to have taken place in the mid 1920s, then $\Delta(\text{Kant}) \cong 1.5$ centuries.

The above historical background may be kept in mind while evaluating the cosmological contributions of Fred Hoyle.

9.3   Interaction of particle physics with cosmology

It is generally assumed that particle physicists and cosmologists first got together in the 1980s, the latter using ideas from particle physics at very high energy in order to address issues like the origin and evolution of large-scale structure. However, the first cosmology to draw heavily on particle physics was the Steady-State cosmology, which explored this frontier area in 1958 at the Paris Symposium on Radio Astronomy. The 'hot universe' of Gold and Hoyle (1959) was the outcome. Briefly, the idea was as follows.

In the Steady-State cosmology, the Universe maintains a steady density despite expansion, by continuous creation of matter. The amount of matter expected to be produced was estimated to be extremely small, at a rate $\sim 10^{-46} \, \mathrm{g \, cm^{-3} \, s^{-1}}$. Nevertheless, the question was, in what form did this new matter appear? Gold and Hoyle proposed the hypothesis that the created matter was in the form of neutrons. The creation of neutrons does not violate any standard conservation laws of particle physics except the constancy of the number of baryons. Although this was considered an objection in 1958, today the number of baryons is no longer regarded as strictly invariant. Indeed, as we shall see later, scenarios based on non-conservation of baryons are being proposed in the context of the very early Universe to account for the observed number of baryons in the Universe.

In the Gold–Hoyle picture the created neutron undergoes a beta decay:

$$n \rightarrow p + e^- + \bar{\nu}. \tag{9.1}$$

The conservation of energy and momentum results in the electron taking up most of the kinetic energy and thereby acquiring a high kinetic temperature of $\sim 10^9$ K. Gold and Hoyle argued that such a high temperature produced inhomogeneously would lead to the working of heat engines between the hot and cold regions, which provide pressure gradients that result in the formation of condensations of size ≥ 50 Mpc. It was already known that pure gravitational forces are not able to provide a satisfactory picture of galaxy formation in an expanding universe. The temperature gradients set up in the hot universe of Gold and Hoyle help in this process.

The resulting system, however, is not a single galaxy, but a supercluster of galaxies containing $\sim 10^3$–10^4 members. Such large-scale inhomogeneities in the distribution of galaxies caution us against applying the cosmological principle too rigorously. For example, if we are in a particular supercluster, we expect to see a preponderance of galaxies of ages similar to that of ours in our neighbourhood out to say 20 or 30 Mpc. Thus it will not be surprising if our local sample yields an average age much larger than the universal average of $(3H_0)^{-1} \approx 3 \times 10^9 h_0^{-1}$ years.

Although newly created electrons have a kinetic temperature of $\sim 10^9$ K, the temperature tends to drop because of expansion. The average temperature is three-fifths of this value, that is, around 6×10^8 K. It was suggested by Hoyle in 1963 that such a hot intergalactic medium would generate the observed X-ray background. However, quantitative estimates by R. J. Gould soon showed that the expected X-ray background in the hot universe would be considerably higher than what is actually observed, thus making the hot universe untenable. The present-day background measurements, however, do not rule out such a hot universe for $h_0 \approx 0.5$. Astrophysicists today are, however, inclined to look for other explanations for the origin of the X-ray background.

Although it is now discredited, the hot universe model was the first exercise in linking particle physics (neutron decay) to the formation of large-scale structures in the Universe.

Notice, however, the difference in approach here from the standard astroparticle physics. The latter relies on untested extrapolation of particle physics coupled with assumed initial conditions for seeding large-scale structure and seeks to arrive at the present hierarchy of structures through several regimes of evolution, neither all directly observable nor analytically calculable. The former takes the process of beta decay, which is well tested in the laboratory and builds on it in timescales of the order of the present-day expansion, to arrive at the supercluster scale structure.

In the 1960s cosmologists by-and-large had not gone beyond classical gravity to address the problem of structure formation; nor had they (as seen in the following section) gone to the extent of accepting structure on the scale of superclusters. The appeal to a particle physics interaction in the above model was therefore viewed with scepticism, and the outcome in the form of superclusters was considered irrelevant to cosmology.

It is somewhat ironical that today cosmologists accept uncritically concepts like GUTs and supersymmetry, a phase transition at 10^{16} GeV, non-baryonic dark matter (cold or hot) as foundations on which to build the evolution of the Universe across a decrease of 87 *orders of magnitude* in density and 29 orders of magnitude in temperature, when *none of the physics of the initial epochs is tested in a laboratory*. Moreover, superclusters are no longer under a taboo, but are well accepted. Thus in this case, we have $\Delta \cong 20$ years for Fred.

9.4   The role of superclusters in radio source counts

In 1961 Martin Ryle and his colleagues at the Mullard Radio Astronomy Observatory in Cambridge announced the results of the 4C radio source survey, claiming that the source counts had a super-Euclidean slope that disproved the Steady-State theory. In a uniform distribution of sources in a Euclidean universe the number N of sources brighter than flux density S goes as $S^{-1.5}$. That is, in the $\log N - \log S$ plot the slope of the number count $N(> S)$ curve will be -1.5. Ryle reported a slope of -1.8, whereas the Steady-State theory was expected to give a slope beginning with -1.5 at high S, and flattening at lower values of S. In January of 1961, Ryle stated these points to claim that the Steady-State theory was disproved.

I had joined as Fred's research student barely six months earlier, and he asked me to develop a counter to Ryle's claim along the following lines:

(a)   Assume that the Universe is inhomogeneous on the scale $\sim$50 Mpc of superclusters. Thus there will be more galaxies in a supercluster, and fewer (ideally zero) in the void outside it.

(b)   Assume that a galaxy becomes a radio source as it ages, i.e. the probability P that the galaxy becomes a radio source increases with age τ. He suggested an empirical formula $P \propto \exp(4H\tau)$.

The supercluster idea had come from the Gold–Hoyle hot universe model. The notion of age-dependence of a radio source property was based on the then-available indications that radio sources do not arise from colliding galaxies but are generally associated with elliptical galaxies (which were considered older than spirals). In any case, Fred Hoyle had maintained the reasonable stand that one should not draw cosmological conclusions from populations of sources whose physics was still unknown. Even today the power-house of a double radio source and the genesis of its jets are hardly well understood.

With these postulates, which in no way altered the basic tenets of the Steady-State cosmology, we were able to demonstrate that an 'average' $\log N - \log S$ curve will have a super-Euclidean slope at high flux levels as found by Ryle *et al.* (Hoyle and Narlikar 1961).

The point that Fred wished to emphasize was that, because of supercluster-scale inhomogeneity, the slope of the $\log N - \log S$ curve fluctuates at large values of S depending on the location of the observer, although at low S it settles down to the cosmological sub-Euclidean value predicted analytically This expectation was later confirmed observationally by deeper surveys (Kellermann and Wall 1987).

To demonstrate this fluctuation, Fred and I thought of carrying out N-body Monte Carlo simulations on an electronic computer. The Cambridge EDSAC was manifestly inadequate for this computation, but Fred had access once a week to an IBM 7090 in London. So with a few weekly visits to London I was able to carry out this demonstration. This was probably the first computer simulation in cosmology (Hoyle and Narlikar 1962).

A great deal was made of the steepness of the $\log N - \log S$ curve at high flux end, with the claim that it implies evolution, which is inconsistent with the Steady-State cosmology. Kellermann and Wall (1987) have commented on how the effect was blown out of proportion, being confined to about 500 relatively nearby sources. Indeed, if the result was cosmologically significant then one must demonstrate that the source population has evolved over the period covered by the survey. For testing evolution one needs to know the redshifts of these sources. Very few redshifts were known in 1961–2. By the mid 1980s, however, most sources in the 3CR catalogue had their redshifts determined. Using this additional information DasGupta *et al.* (1988) were able to show that no evolution was necessary for the consistency of most Friedmann models (with $\Lambda = 0$), with the source-count data as per the 3CR catalogue. DasGupta (1988) later showed also that even the Steady-State cosmology was consistent with the 3CR source count. Similar complete redshift data for the 4C survey are not yet available for carrying out such analysis.

In the 1960s the concept of superclusters was not 'standard', and most cosmologists believed that the Universe was homogeneous on scales larger than clusters of galaxies ($\sim$5 Mpc). The idea that the Universe can be inhomogeneous on the supercluster scale introduces a larger degree of fluctuations in the predicted values of observational tests of homogeneous cosmology. Evidence existed from the studies of Abell (1958), de Vaucouleurs (1961), Shane and Wirtanen (1954) on superclusters, but nobody believed that the Universe could be inhomogeneous on such a large scale. The 'complication' introduced by inhomogeneity on the scale of superclusters ($\sim$50 Mpc) was therefore felt unnecessary in the opinion of many theoreticians, and certainly a high price to pay in order to keep the Steady-State theory alive. It was some two decades later, in the 1980s, that the existence of superclusters and voids on scales of 50–100 Mpc became part of standard cosmology. Thus in this instance, I would set $\Delta \cong 25$ years for Hoyle's belief in superclusters.

I now return to the interaction between cosmology and particle physics.

9.5   Non-conservation of baryons and negative stress energy

I understand that Fred had sent his first manuscript on the Steady-State cosmology to a well known physics journal. It was rejected there presumably because physicists looked upon continuous creation as a violation of the law of conservation of matter and energy. (The reason for rejection cited by the journal, however, was a curious one, namely that it was facing shortage of paper.) He subsequently sent it to the astronomy journal *MNRAS*. In fact, unlike the version of Bondi and Gold (1948), the version of Steady-State cosmology advocated by Fred Hoyle (1948) *does not* violate the above conservation law. There, a scalar field of negative energy and pressure was used, an idea that physicists found abhorrent. It is significant that the idea is now gaining popularity, see its recent 'rediscovery' by Steinhardt and Turok (2002). Thus one could argue that $\Delta \cong 50$ years for this idea originally proposed by Hoyle.

The Gold–Hoyle hot universe model had continuous creation of neutrons. In general Hoyle believed that baryons (in preference to antibaryons) would be created. This breaks the baryon-number conservation law as well as baryon–antibaryon symmetry, which were considered sacrosanct in the 1960s. Thus when our paper (Hoyle and Narlikar 1966a) on non-conservation of baryons in cosmology came up the physicists who took note of it argued that the idea violated the above principles.

Again it is significant that with the approach to Grand Unified Theories particle physicists themselves found these principles no longer necessary. Indeed they were highly constraining to Big-Bang cosmology if one wished to explain the observed baryon–antibaryon asymmetry and the baryon to photon ratio. In the end, high-energy particle physicists have dropped these symmetries.

On one occasion Fred Hoyle himself answered the criticism on baryon non-conservation by stating that this is the consequence of broken symmetry which perpetuates itself. The *C*-field which mediates in the creation process may have internal degrees of freedom that favour matter over antimatter. Since in later (post-1964) versions of the *C*-field, action at a distance formulation was used (see Hoyle and Narlikar 1964), one could argue that the information of broken symmetry in one spacetime event could be carried along light cones to the future and thus spread all over the Universe.

If we date the notion of baryon non-conservation in cosmology to the Hoyle–Narlikar paper of 1966, and look at the 1979 publication by Steven Weinberg (entitled 'Baryon-lepton non-conserving processes'), we may set $\Delta = 13$ years for this idea. In fact all three of the trilogy of papers published by Hoyle and me in 1966 have found echoes in subsequent years as we shall see in the following two sections.

9.6   Inflation and the bubble universe

I now come to the field theory with which Hoyle and I worked in order to derive the physical properties of the Steady-State universe related to gravity and matter creation. The C-field theory, as it is called, was in fact based on the scalar-field formulation provided by M. H. L. Pryce in 1961 as a private communication.

Like Hoyle's original approach, the C-field theory also involved adding more terms to the standard relativistic Einstein–Hilbert action to represent the phenomenon of creation of matter. Using Occam's razor, the additional field to be introduced was a scalar field with zero mass and zero charge. We denote this field by C and its derivative with respect to the spacetime coordinate x^i by C_i. The action is then given by (with $c = $ speed of light),

$$\mathcal{A} = \frac{c^3}{16\pi G} \int R\sqrt{-g}\, d^4x - \Sigma_a m_a c \int ds_a$$
$$- \frac{1}{2c} f \int \int C_i C^i \sqrt{-g}\, d^4x - \Sigma_a \int C_i da^i. \tag{9.2}$$

The additional terms (third and fourth) on the right-hand side are the C-field terms. Note that the last term of (9.2) is path-independent. If we consider the world line of particle a between the end points A_1 and A_2, we have

$$\int_{A_1}^{A_2} C_i\, da^i = C(A_2) - C(A_1). \tag{9.3}$$

Normally such path-independent terms do not contribute to any physics derivable from the action principle. So why include such a term? The answer to this question lies in the notion of 'broken' world lines. A theory that discusses creation (or annihilation) of matter per se must have world lines with finite beginnings or ends (or both). The C-field interaction term picks out precisely these end points of particle world lines. If we vary the world lines of a and consider the change in the action $\mathcal{A}$ in a volume containing the point A_1 where the world line begins, we get at A_1 (which is now varied)

$$m_a c \frac{da^i}{ds_a} g_{ik} - C_k = 0. \tag{9.4}$$

This relation tells us that *overall energy and momentum are conserved at the point of creation*. The 4-momentum of the created particle is compensated by the 4-momentum of the C-field. Clearly, to achieve this balance the C-field must have negative energy. We shall return to this point later. We also note that, since the interaction term is path-independent, the equation of motion of a is still that of a geodesic. The Pryce formulation is therefore a masterly way of dealing with creation (and annihilation) of matter without violating the conservation laws.

The constant f in the action (9.2) is a coupling constant. The variation of C gives the source equation in the form

$$C_{ik}^{\ k} = c f^{-1} \bar{n}, \tag{9.5}$$

where $\bar{n}$ is the number of net creation events per unit proper 4-volume.

Finally, the variation of g_{ik} leads to the modified Einstein field equations

$$R^{ik} - \frac{1}{2}g^{ik}R = -\frac{8\pi G}{c^4}\left(T_{(m)}^{ik} + T_{(C)}^{ik}\right),$$
(9.6)

where $T_{(m)}^{ik}$ is the matter tensor while

$$T_{(C)}^{ik} = -f\left(C^i C^k - \frac{1}{2}g^{ik}C^l C_l\right).$$
(9.7)

Again we note that $T_{(C)}^{00} < 0$ for $f > 0$. Thus the C-field has a negative energy density that produces a repulsive gravitational effect. It is this repulsive force that drives the expansion of the Universe. The above effect may resolve one difficulty usually associated with the quantum theory of negative energy fields. Because such fields have no lowest energy state, they normally do not form stable systems. A cascading into lower and lower energy states would inevitably occur if we perturb the field in a given state of negative energy. However, this conclusion is altered if we include the feedback of (9.7) on spacetime geometry through (9.6). This feedback results in the expansion of space and in the lowering of the magnitude of field energy. These two effects tend to work in opposite directions and help stabilize the system.

Using the Robertson–Walker line element and the assumption that a typical particle created by the C-field has mass m, we get the following equations out of the above set:

$$\dot{C} = mc^2,$$
(9.8)

$$mf\left(\ddot{C} + 3\frac{\dot{S}}{S}\dot{C}\right) = \left(\dot{\rho} + \frac{\dot{S}}{S}\rho\right)c^2,$$
(9.9)

$$2\frac{\ddot{S}}{S} + \frac{\dot{S} + kc^2}{S^2} = \frac{4\pi G f}{c^4}\dot{C}^2,$$
(9.10)

$$3\frac{\dot{S}^2 + kc^2}{S^2} = 8\pi G\left(\rho - \frac{f}{2c^4}\dot{C}^2\right).$$
(9.11)

It is easy to verify that the steady-state solution follows from these equations for

$$k = 0, \qquad S = e^{H_0 t}, \qquad \rho = \rho_0 = \frac{3H_0^2}{4\pi G} = fm^2.$$
(9.12)

Notice that both H_0 and ρ_0 are given in terms of the elementary creation process: that is, in terms of the coupling constant f and the mass of the particle created. Thus the Hoyle approach provides the quantitative information lacking in the deductive approach via the Perfect Cosmological Principle of Bondi and Gold.

A first-order perturbation of the above equations and of the steady-state solution also tells us that the solution is stable. Indeed, a stability analysis brings out the key role played by the creation process. This tells us that the created particles have their world lines along the normals to the surfaces $C = $ constant. Hoyle has argued that such a result gives a physical justification for the Weyl postulate; it tells us *why* the world lines of the fundamental observers are orthogonal to a special family of spacelike hypersurfaces. In the C-field cosmology these hypersurfaces are not just abstract notions but are seen to have a physical basis. We therefore argued that even if the Universe was considerably different from the homogeneous and isotropic form in the remote past, the creation process would drive it to that state eventually. Years later this idea resurfaced in the context of inflation as the 'cosmic no-hair conjecture', namely that an inflationary universe wipes out the initial irregularities and leads to homogeneity and isotropy. It has been recognized by Barrow and Stein Schabes (1984) that this notion is very similar to the above result derived by us in the early sixties (Hoyle and Narlikar 1963; $\Delta = 21$ years!).

However, as it turned out, Fred had anticipated the very idea of inflation in the mid 1960s. This was published in a paper with myself as coauthor (Hoyle and Narlikar 1966b) where we discussed the effect of raising the coupling constant f by $\sim 10^{20}$. As the formulae (9.12) show, we would then have a Steady-State universe of very large density ($\rho_0 \simeq 10^{-8}$ g cm^{-3}) and very short timescale ($H_0^{-1} \simeq$ 1 year!). If in such a dense universe creation is switched off in a local region, that is, if we locally have a phase transition from the creative to the non-creative mode:

$$C^i_{;i} = 0, \tag{9.13}$$

then this local region will expand according to the formula

$$S(t) \propto \left[1 + \frac{(t + t_1)^2}{t_0^2} \right]^{1/3}, \tag{9.14}$$

where t_1 and t_0 are constants. Note that this is the 'non-singular' analogue of the Einstein–de Sitter model of standard cosmology (now more popularly known by the parameters $\Omega_{\text{matter}} = 1$, $\Omega_\Lambda = 0$), which has $S(t) \propto t^{2/3}$. Indeed, for small t_0, the solution rapidly approaches the Einstein–de Sitter form. Being less dense than the surroundings, such a region will simulate an air bubble in water. Although the basic physics is different, the similarity between this model and the inflationary model that came into fashion 15 years later is obvious. In both models a phase transition creates the bubble that expands into the outer de Sitter spacetime. In the Steady-State universe, such bubbles could arise in many places at different epochs from $t = -\infty$ to $t = +\infty$.

According to this model, this bubble is all that we see with our surveys of galaxies, quasars and so on. Hence our observations tell us more about this unsteady perturbation than about the ambient Steady-State universe. There are, however, observable effects that give indications of the high value of f. For example, we showed that particle creation is enhanced near already existing massive objects and that the resulting energy spectrum of the particles would simulate that of high-energy cosmic rays. The actual energy density of cosmic rays requires the high value of f chosen here.

Thus taking Fred's anticipation of inflation in 1966, we may set $\Delta = 15$ years.

9.7   Nuclei of galaxies

The following extract from the abstract of the Hoyle and Narlikar (1966c) paper will indicate Fred's ideas in the mid 1960s on the dynamics of galaxy formation:

> We suggest that the condensation of...galaxies depends on the
> presence of inhomogeneities, in particular that a galaxy is formed
> around a central mass concentration. Because the Einstein–de Sitter
> expansion law is the limiting case between the expansion to infinity at
> finite velocity and a fall-back situation, in which the expansion stops
> at some minimum but finite density, a central condensation with
> mass appreciably less than that of the associated galaxy suffices to
> prevent continuing expansion. A mass of $10^9 \, M_\odot$, for example, will
> restrain a total mass of $\sim 10^{12} \, M_\odot$ from expanding beyond normal
> galactic dimensions...

In the mid 1960s the notion of a massive black hole at the nucleus of a galaxy had not received 'standard sanction', and so the idea remained relatively unknown, especially because it was proposed in the context of a Steady-State universe. I briefly elaborate on the idea that the above abstract indicates, while stressing that the arguments were made in the mid 1960s.

The cosmological basis of this work was discussed in the preceding paper (Hoyle and Narlikar 1966b), which supposed that the Universe, or a portion of it, expands from an initially steady-state situation with $\rho \simeq 10^{-8} \, \mathrm{g \, cm^{-3}}$, $H^{-1} \simeq 10^{18}$ cm, that creation is effectively zero during this expansion, and that the Einstein–de Sitter expansion law holds in first approximation.

The Newtonian analogue of the Einstein–de Sitter law is given by

$$\dot{r}^2 = 2GM/r \tag{9.15}$$

in which r is the radial coordinate of an element of material defined by the condition that in a spherically symmetric situation about $r = 0$, the mass interior to r is M. For a given sample of material M remains constant and $\dot{r} \to 0$ only as $r \to \infty$. Equation (9.15) is an integral of the second-order Newtonian equations, and the fact that no constant of integration appears represents the analogue of the Einstein–de Sitter law.

Next, consider the Newtonian problem of an object of mass μ placed at the origin $r = 0$, all conditions at a particular moment for a particular element of the cloud being the same as before. Denote the value of r at this moment by r_0. Then $\dot{r}$ at this moment is $(2GM/r_0)^{1/2}$, as before, and the subsequent motion of the element in question is determined by

$$\dot{r} = \frac{2G(M + \mu)}{r} - \frac{2G\mu}{r_0}. \tag{9.16}$$

The outward velocity drops to zero, and the element subsequently falls back towards $r = 0$. The maximum radial distance r_{max} reached by the element is given by

$$r_{max} = \{1 + (M/\mu)\}r_0, \tag{9.17}$$

and for sufficiently large M/μ, $r_{max} \simeq Mr_0/\mu$, so that the fractional increase r_{max}/r_0, above the radius r_0 at which the element had the same radial motion as in the Einstein–de Sitter case, is just M/μ. This factor is larger for elements more distant from μ than for the inner parts of the cloud, so the outer parts recede proportionately further than the inner parts.

What determines the particular moment at which the Einstein–de Sitter condition, $\dot{r} = (2GM/r)^{1/2}$, holds for any particular sample of material? To come to grips with this important question we must consider the relativistic formulation of the problem.

A complete solution of a local gravitational problem can be represented as a power series in the dimensionless parameter $2G(M + \mu)/r$, which must be $\ll 1$, this being what we mean by a 'local problem'. The Newtonian solution is of course the first term in this series. However, it is clear that we cannot use the Newtonian solution for the effect of μ if the second-order term in $2GM/r$ exceeds the first-order term in $2G\mu/r$, as is possible when $\mu/M \ll 1$. Hence the Newtonian equations for the effect of μ, namely equations (9.16) and (9.17), cannot be used unless the moment for which we use $r \equiv r_0$, $\dot{r} = (2GM/r_0)^{1/2}$, is such that

$$\frac{2G\mu}{r_0} \geq \left(\frac{2GM}{r_0}\right)^2. \tag{9.18}$$

By taking equality in (9.18) we do indeed define a particular value of r, corresponding to a specified M, namely,

$$r_0 = 2GM \times \left(\frac{M}{\mu}\right). \tag{9.19}$$

The situation is that the Newtonian calculation for the effect of μ can be applied to the subsequent motion of an element of material such that the specified M lies interior to it. But can we use $(2GM/r_0)^{\frac{1}{2}}$ as the starting velocity in this calculation? Not in general, because in general the cloud will have at least small fluctuations from the Einstein–de Sitter expansion. We shall confine ourselves here to the case in which the conditions $r \approx r_0$, $\dot{r} = (2GM/r_0)^{\frac{1}{2}}$, with r_0 given by equation (9.19), hold for all M.

From equations (9.17) and (9.19) we have

$$r_{max} \simeq \frac{M}{\mu} r_0 \simeq 2GM \left(\frac{M}{\mu}\right)^2. \tag{9.20}$$

This result has a number of interesting consequences. Set r_{max} equal to a typical galactic radius $r_{max} = 3 \times 10^{22}$ cm. Then equation (9.20) leads to

$$\frac{M}{M_\odot} \simeq 5 \times 10^5 \left(\frac{\mu}{M_\odot}\right)^{\frac{2}{3}}, \tag{9.21}$$

where $M_\odot$ is the solar mass. A central object of mass $\mu = 10^9 \, M_\odot$ gives $M = 5 \times 10^{11} \, M_\odot$, while $\mu = 10^7 \, M_\odot$ gives $M = 2 \times 10^{10} \, M_\odot$. It is of interest that the central condensations present in massive elliptical galaxies are known to be of order $10^9 \, M_\odot$, and that the total masses are believed to be $\sim 10^{12} \, M_\odot$.

Suppose that during expansion stars are formed from gas. The stars will continue to occupy the full volume corresponding to their maximum extension from the centre, so that the mass of the stars interior to r is given by setting $r_{max} = r$ in equation (9.21). Numerically, we have

$$\frac{M(r)}{M_\odot} \simeq 2 \times 10^5 \left(\frac{\mu}{M_\odot}\right)^{\frac{2}{3}} r^{\frac{1}{3}}, \tag{9.22}$$

in which r is in kiloparsecs. Evidently, the mean star density at distance r from the centre is proportional to M/r^3, i.e., to $r^{-\frac{8}{3}}$. So long as the stars have everywhere the same luminosity function, the emissivity per unit volume at distance r is proportional to $r^{-\frac{8}{3}}$. This determines the light distribution in a spherical elliptical galaxy.

To obtain the projected intensity distribution we first note that the above considerations can be applied to values of r beyond normal galactic dimensions. There is no upper limit to r so long as we are dealing with a single condensation. This agrees with observation, in that no ultimate maximum radius has yet been found; the conventional radii are simply those set by the sensitivity of particular observing techniques. This being so, the intensity distribution, $I(r)$, of the projected image is obtained by multiplying the volume emissivity by the factor r, and is $I(r) \propto r^{\frac{5}{3}}$. This proportionality is slightly less steep than Hubble's law for $r \gg a$:

$$I(r) \propto (r/a + 1)^{-2} \approx r^{-2}. \tag{9.23}$$

The measurements for early ellipticals E1, E2, E3 give very good agreement with $r^{-\frac{5}{3}}$, better than with r^{-2}.

The $r^{-\frac{5}{3}}$ proportionality must not be used for too-small r, because r_0 given by equation (9.19) becomes invalid as M is reduced towards μ. The reason is simply that if M is set too small the mean density corresponding to equation (9.19), namely $M/\frac{4}{3}\pi r_0^3 \propto M^{-5}$, becomes larger than the steady-state value of $\sim 10^{-8}\, \mathrm{g\, cm^{-3}}$ from which the expansion started. Instead of equation (9.19), we then have an initial radius r_i given by

$$\frac{4}{3}\pi r_i^3 \rho_i = M, \qquad \rho_i \simeq 10^{-8}\, \mathrm{g\, cm^{-3}}, \tag{9.24}$$

and instead of equation (9.20),

$$r_{\max} \simeq \frac{M}{\mu} r_i, \qquad \frac{M}{\mu} \gg 1. \tag{9.25}$$

In place of equation (9.22) we have

$$r \simeq 10^{-5} \frac{M}{\mu}\left(\frac{M}{M_\odot}\right)^{\frac{1}{3}}, \tag{9.26}$$

with r now in parsecs. As an example, for $M = 10^{11}\, M_\odot$, $\mu = 10^8\, M_\odot$, equation (9.26) gives $r \simeq 30$ parsecs. This result is very satisfactory in that it predicts a highly concentrated point of light at the centres of elliptical galaxies.

Note that in this scenario the formation of a massive object in the centre of the galaxy is not discussed. In 1965–6, Fred and I assumed its existence and worked out consequences of the above type. The creation process is expected to generate more mass preferentially near an existing massive object, and so the mass grows to a large size, until the accumulation of the excess C-field leads to repulsive instabilities, and explosive phenomena may occur. This process was discussed in detail in the book by Hoyle, Burbidge and Narlikar (2000).

This idea too was not paid much attention to by those interested in the cosmogony of galaxies, partly because the standard methods of gravitational contraction of gas clouds could not give such collapsed objects as endstates. To-day, however, there is a great enthusiasm for the existence of supermassive black holes in the nuclei of galaxies, both in terms of their theoretical consequences and the observational features. Thus again we find Fred ahead of the pack by $\Delta \cong 20$ years.

However, in the standard scenario the formation of a supermassive object through gravitational collapse is still not properly understood, the reason being the same as that which led to the general scepticism of the concept in the 1960s. A major difficulty is how to get rid of the angular momentum of the initial state from which collapse is supposed to ensue.

Observation does not show the presence of massive black holes in the *discs* of spiral galaxies. Molecular clouds with masses upwards of $10^5 \, M_\odot$ are found, but only with average matter densities of $\sim 10^{-21} \, g \, cm^{-3}$. The reason why there is no continuing condensation of such clouds is usually put down to rotary forces. When a cloud condenses, the inward gravitational forces increase as the inverse square of the scale of the cloud. But the rotary forces inhibiting condensation increase as the cube. Since the two are not far from being in balance initially, not much condensation is permitted before rotation becomes inimical to any further rise of the internal density.

The internal density inside a black hole of mass $10^6 \, M_\odot$, say, is about $10^4 \, g \, cm^{-3}$. So a molecular cloud of initial average density $10 - 21 \, g \, cm^{-3}$ would have to condense by upwards of a factor 10^8 in scale to produce a black hole, with rotary forces increasing by the immense factor of $\sim 10^{25}$. No amount of optimistic thinking can cope with an increase of that order. So what now is special about the centre of a galaxy, instead of the disc, to permit a rise of rotary forces by this same enormous factor? Nothing. The centre of a galaxy is certainly a unique point geometrically. But rotary forces are not suspended there. The situation is just the same there as for the discs of spirals. The alternative picture involving creation of matter described above does not have this problem.

9.8   Is the universe accelerating?

Recently there has been considerable hype on the 'accelerating universe'. The source of this enthusiasm for the accelerating models is in the observations of redshifts z and apparent magnitudes m of distant (high redshift) supernovae. In the expanding universe model, the apparent magnitude can be related to the redshift through an explicit relation that depends on the model chosen, provided (1) the light source used (in this case the peak luminosity of the supernova) is truly a standard candle, and (ii) there is no intergalactic absorption en route from the source to the observer.

In the 1960s and 1970s, Allan Sandage and his collaborators played an extensive role in applying this test to the expanding-universe models. At the time, the invariable conclusion from such studies was that the Universe is *decelerating*. Indeed, standard texts in cosmology usually define a *deceleration parameter* q_0 by

$$q_0 = -\frac{\ddot{S}}{S} H_0^{-2}, \tag{9.27}$$

where H_0 is the present value of the Hubble constant. Sandage usually quoted values of this parameter ranging from 1 down to almost zero, but positive. All Friedmann models then under discussion had $\Lambda = 0$ and predicted positive q_0.

There was one joker in the pack, though! The Steady-State model with $S \propto \exp(Ht)$ predicted $q_0 = -1$. It was singled out as an example of a wrong cosmology.

Today the situation is the other way round: the general consensus is that q_0 is negative. However, I am disappointed to see that none of the experimental groups associated with this result have made a reference to the Steady-State theory as giving the right value of q_0. The Steady-State theory may be faulted on other counts, but surely it does deserve a pat on the back for its prediction of an accelerating universe.

Why does the Steady-State theory predict an accelerating universe? This is because it employs, in Fred Hoyle's approach, a *negative-energy scalar field*, namely, the C-field. A negative-energy field used in Einstein's equations produces repulsion, and hence acceleration. Today, attempts are being made to put dynamics behind the Λ-term, with claims of quintessence or dark energy being already made with the same degree of firmness and confidence as was found in the Agnes Clerke quotation in Section 9.2. As is apparent from the work of Steinhardt and Turok (2002), a consensus will eventually develop that this effect is possible only under the regime of a negative-energy field. But again hardly anyone would bother to reference the work on the C-field which precedes the present work by four decades.

However, let me return to the two provisos mentioned at the beginning of this section. Are we sure that we are dealing with standard candles? Recall that much of the work in the 1960s and 1970s involving galaxies as standard candles was vitiated by the possibility that there might be luminosity evolution in galaxies. How sure are we that a supernova at $z = 1.6$ has the same peak luminosity as a present-day supernova? A second element of uncertainty introduced recently with regard to high redshift supernovae comes from gravitational lensing (Moertsell *et al.* 2001). At high redshift there is greater chance of a supernova being gravitationally lensed in such a way as to amplify its luminosity. The second proviso has been highlighted by Aguirre (1999), Banerjee *et al.* (2000) and Narlikar *et al.* (2002a), by pointing out that intergalactic dust can produce extra dimming of distant supernovae and thus apparently simulate the effect of a positive cosmological constant.

Recently Narlikar *et al.* (2002a) have argued that the quasi-steady-state cosmology (QSSC) (which employs a *negative* cosmological constant) produces an m–z relation for supernovae that is fully consistent with observations including that of the high-redshift supernova 1997ff. Here the creation field used by the QSSC behaves like a positive cosmological constant (as in the Steady-State theory); however, the main effect is produced by the intergalactic dust. It is significant that the magnitude of the dust density required for thermalizing starlight in order to generate the cosmic microwave background radiation (CMBR) in the QSSC is fully consistent with the value obtained for a good fit of the theoretical m–z curve to the observations.

Intergalactic dust is another concept that Fred proposed back in the 1970s in order to explain the CMBR as thermalized starlight. It was in the 1990s, within the framework of the QSSC, that the idea found a workable framework (Hoyle *et al.* 2000). For it is not only possible to demonstrate that the starlight from stars of previous generations can be adequately thermalized by such dust, but one also gets the present-day temperature of CMBR as 2.7 K, a feat not yet achieved by standard cosmology. Further, as shown by Narlikar *et al.* (2002b), one can also understand the angular power spectrum of inhomogeneities of the CMBR.

9.9   A general comment

I have given these instances to counter the impression generally created that Fred was right about stellar evolution, nucleosynthesis, and molecular astronomy but mostly wrong about cosmology. His perception of large-scale inhomogeneity of the Universe on the supercluster scale, the use of Monte Carlo N-body simulations in cosmology, his appreciation of a possible role that particle physics could play in cosmology, the bold assertion that the baryon number is not conserved, the anticipation of a model very similar to that of inflation, the inclusion of negative-energy–negative-stress fields in the dynamics of the Universe and the notion that galaxies have compact massive nuclei controlling their dynamics and shapes were regarded as outlandish at the time they were

proposed, but became part of mainstream cosmology when proposed by others much later.

It is unfortunate that later generations 'rediscovering' these ideas have either been ignorant of Fred's earlier work or have been aware of it but have chosen to ignore it. Examples of the former are to be found in the papers by Guth (1981) and Linde (1987), and of the latter in the recent work of Steinhardt and Turok (2002).

The cartoons (1)–(3) illustrate three kinds of interaction an individual scientist may have vis-à-vis mainstream research, representing a man catching a bus named (appropriately) the 'Bandwagon'. Cartoon (1) shows a typical bright young scientist who is wise enough to base his research on mainstream ideas, for that way lies progress, promotion and prosperity. He gets on the bus at the right time. Cartoon (2) represents a scientist who has thought of an idea too late, for it is already known to the community: he has rightly missed the bus. Cartoon (3) shows a scientist like Fred Hoyle who was years ahead of his times. The bandwagon follows him, but alas, far from giving him the credit for his ideas, knocks him out!

Acknowledgement

I thank Douglas Gough for travel support which made my participation in the Hoyle meeting possible. For the cartoons I am indebted to my wife Mangala.

References

ABELL, G. O. 1958 ApJS, **3**, 211

AGUIRRE, A. N. 1999 ApJ, **512**, L19

BANERJEE, S. K., NARLIKAR, J. V., WICKRAMASINGHE, N. C., HOYLE, F. & BURBIDGE, G. 2000 AJ, **119**, 2583

BARROW, J. D. & STEIN SCHABES, J. 1984 Phys. Lett., **103A**, 315

BONDI, H. & GOLD, T. 1948 MNRAS, **108**, 252

CLERKE, A. 1905 The System of the Stars (London: Adam & Charles Black) p. 349

DASGUPTA, P. 1988 Ph.D. Thesis, Bombay University

DASGUPTA, P., NARLIKAR, J. V. & BURBIDGE, G. 1988 AJ, **95**, 5

DISNEY, M. 2000 GRG Journal, **32**, 1125

GOLD, T. & HOYLE, F. 1959 Proceedings of the Paris Symposium on Radio Astronomy, ed. R. N. Bracewell (Stanford: Stanford University Press) p. 583

GUTH, A. 1981 Phys. Rev. D, **23**, 347

HOYLE, F. 1948 MNRAS, **108**, 372

HOYLE, F. & NARLIKAR, J. V. 1961 MNRAS, **123**, 133

1962 MNRAS, **125**, 13

1963 Proc. Roy. Soc. A, **273**, 1

1964 Proc. Roy. Soc. A, **282**, 178

1966a Proc. Roy. Soc. A, **290**, 143

1966b Proc. Roy. Soc. A, **290**, 162

1966c Proc. Roy. Soc. A, **290**, 177

HOYLE, F., BURBIDGE, G. & NARLIKAR, J. V. 2000 A Different Approach to Cosmology (Cambridge: Cambridge University Press)

HUBBLE, E. P. 1926 ApJ, **63**, 236

KELLERMANN, K. & WALL, J. V. 1987 Proceedings IAU Symposium No. 124, eds. A. Hewitt, G. Burbidge & L.-Z. Fang (Dordrecht: Reidel) p. 545

LINDE, A. 1987 Physica Scripta, **T15**, 169

VAN MAANEN, A. 1916–30 Mt Wilson Contr., Nos. 111, 136, 158, 182, 204, 237, 270, 290, 321, 356, 391, 405–408

MOERTSELL, E., GUNNARSSON, C. & GOOBAR, A. 2001 ApJ, **561**, 106

NARLIKAR, J. V. & KEMBHAVI, A. K. 1980 Fund. Cosmic Phys., **6**, 1

NARLIKAR, J. V. & PADMANABHAN, T. 2001 *ARA&A*, **39**, 211

NARLIKAR, J. V., VISHWAKARMA, R. G. & BURBIDGE, G. 2002a *PASP*, **114**, 1092

NARLIKAR, J. V., VISHWAKARMA, R. G., HAAJIAN, A., SOURADEEP, T., BURBIDGE, G. & HOYLE, F. 2002b *ApJ*, **585**, 1

PENZIAS, A. A. & WILSON, R. W. 1965 *ApJ*, **142**, 419

SHANE, C. D. & WIRTANEN, C. A. 1954 *AJ*, **59**, 285

STEINHARDT, P. J. TUROK, N. 2002 *Science*, **296**, 1436

DE VAUCOULEURS, G. 1961 *AJ*, **66**, 629

WEINBERG, S. 1979 *Phys. Rev. Lett.*, **43**, 1566

10

Red giants: then and now

JOHN FAULKNER

UCO/Lick Observatory, University of California Santa Cruz

Fred Hoyle's work on the structure and evolution of red giants, particularly his pathbreaking contribution with Martin Schwarzschild (Hoyle and Schwarzschild 1955), is both lauded and critically assessed. In his later lectures and work with students in the early 1960s, Hoyle presented more physical ways of understanding some of the approximations used, and results obtained, in that seminal paper. Although later ideas by other investigators will be touched upon, Hoyle's viewpoint – that low-mass red giants are essentially white dwarfs with a serious mass-storage problem – is still extremely fruitful. Over the years, I have further developed his method of attack. Relatively recently, I have been able to deepen and broaden the approach, finally extending the theory to provide a unifying treatment of the structure of low-mass stars from the main sequence though both the red-giant and horizontal-branch phases of evolution. Many aspects of these stars that had remained puzzling, even mysterious, for decades have now fallen into place, and some questions have been answered that were not even posed before.

With low-mass red giants as the simplest example, this recent work emphasizes that stars, in general, may have at least *two* distinct but very important centres: (i) a *geometrical* centre, and (ii) a separate *nuclear* centre, residing in a shell outside a zero-luminosity dense core for example. This two-centre perspective leads to an explicit, analytical, asymptotic theory of low-mass red-giant structure. It enables one to appreciate that the problem of understanding why such stars become red giants is one of anticipating a remarkable yet natural structural bifurcation that occurs in them. This bifurcation occurs because of a combination of known and understandable

The Scientific Legacy of Fred Hoyle, ed. D. Gough.
Published by Cambridge University Press. © Cambridge University Press 2004.

facts just summarized: namely that, following central hydrogen exhaustion, a thin nuclear-burning shell *does* develop outside a more-or-less dense core.

In the resulting theory, both $\rho_{sh}/\bar{\rho}_c$ and $\rho_{sh} \cdot \bar{\rho}_c$ prove to be important self-consistently derived quantities. I present some striking, *explicit*, asymptotic analytical theorems and results involving these quantities. Perhaps the most astonishingly unexpected and gratifying single result is this: for the very value Nature gives us for the relevant temperature exponent ($\eta = 15$; CNO cycle) for nuclear-energy generation, ρ_{sh} and $\bar{\rho}_c$ behave in a well defined, *precisely inverse* manner for a given value of core-mass, M_c. This emphasizes that the internal behaviour of such stars is definitely *anti-homologous* rather than *homologous*: dense cores physically promote diffuse surrounding envelopes. I also extend the ideas yet further in a way which (i) links the structural and evolutionary behaviour of stars from the main sequence through horizontal-branch phases of evolution, and (ii) also has implications for post-main-sequence developments in more massive stars.

The end result is that the post-main-sequence developments of *all* stars – low-mass, intermediate-mass, and high-mass – as they expand to become giants, are finally seen to be examples of one underpinning fact: that dense cores with thin surrounding shells naturally follow hydrogen exhaustion. While 'this has been know all along' from oft-repeated computer calculations, we now know why analytically. That matters to true theorists.

What follows is a requested, much expanded version of my Cambridge talk.

10.1   Introduction

The memorial meeting was in many ways a sad occasion for us all. Nevertheless, it was a distinct personal pleasure to be in Cambridge, at such a special meeting celebrating Fred Hoyle's work and honouring his memory. I am very grateful to Douglas Gough for his invitation to speak at the meeting and then write about 'Red giants, then and now'. The unusual structure and behaviour of evolving red giants were subjects always close to Fred's heart. This chapter gives me not only the opportunity to discuss his pioneering red giant ideas and computations, but also to describe something of how his own thinking on this topic developed in later years, beyond that in his published work. I shall also discuss further clarifying conclusions that I have been able to bring to Fred's insightful approach.

My contribution will have two main themes. First of all, I shall write about Fred's own published work on understanding the structure and evolution of red giants, particularly his great 1955 paper with Martin Schwarzschild – truly one of the most significant landmark papers of the last half century. I shall precede that by a discussion of the highlights of some of his earlier work, because that helps to set the astrophysical scene at the time, as well as providing the logical lead-up to this truly seminal paper. In addition, I shall describe some of his subsequent work with Brian Haselgrove, on dating the globular clusters, etc., as well as some of the difficulties the two encountered in proceeding further in understanding the horizontal-branch stars. I consider this all part of the heroic early days of computational stellar evolution. I shall call this section, 'Standing on the shoulders of giants!'

That model stars *do* become [model] red giants, following their structurally much simpler and well understood main-sequence phases, has become a computational commonplace, and something that 'everybody knows'. However, that 'known fact' does not (or at least, *should not*) satisfy the true astrophysical theorist, who always wants to know 'Why?' of course. The true theorist isn't satisfied by the answer: 'Every time we compute models, that's what happens'.[1] So the key theoretical question is: 'Is there a particular, significant physical circumstance that one can pinpoint as the reason for red giants becoming so different and behaving so differently from their forbears?' There have been many (*very* many) theoretical attempts to 'understand' or 'explain' why stars become red giants. I shall give a nod to some of them. But for my money, I still take a viewpoint that Fred explained in lectures, but which was never fully published, as the best way of understanding the structure of red giants, why it's so different from that of main-sequence stars, and thus why stars *must* become red giants. That will be the theme of the later part of this chapter.

I was most fortunate to have been exposed as a student to Fred Hoyle's thinking on this subject, perhaps more intimately than anyone else. However, at that time he and I took it only to a limited end, namely that of understanding the strong dependence of giants' luminosities on their core-masses. More recently, I've been able to identify and understand a key component of the argument much more deeply, and thus to bring Fred's almost intuitive approach to full fruition, in a way that some find remarkably simple yet utterly compelling. Many structural consequences follow from this, including two asymptotic theorems explicitly demonstrating the necessity of both high luminosities *and* low shell, and thence low envelope, densities in such stars. The latter low-density requirement,

---

[1] This point, while appropriate for what I consider true theorists to be, may not necessarily hold for today's vast legions of large numerical simulators, many of whom seem so strongly focused on numerical reproduction. One must question whether this is 'understanding' when, as happens quite often, the consequences of a change in just one key parameter cannot be foretold. Oddly enough, I have noticed an interesting thing about human beings when they are engaged in activities that can lead to reproduction: careful analytical thinking about serious intellectual questions is often the last thing on their minds.

self-consistently derived from and coupled with a critical-temperature condition, is what brings about the extremely large radii of red giants. (I have also been able to extend Fred's approach to the treatment of horizontal-branch stars. By doing that, I have been able to answer some questions that no one, including myself, seems to have posed before, and to throw new light on my own original work (Faulkner 1966) in first exploring in a systematic way the structure and parameter dependences of such stars.)

These results finally answer that decades-old question (and more): 'Why and how do stars become red giants?' At least, I was honoured and delighted to receive Hermann Bondi's enthusiastic and spontaneous assertion to that effect, immediately following my talk in Cambridge. I was tempted to call this later part of the chapter: 'Standing on the shoulders of dwarfs!' – not because I wish to cast aspersions on all the many theorists who have worked on this problem (heaven forfend), but rather because, as we shall see, it summarizes the very foundation of Fred's approach in a brief and aphoristic way.

10.2   Standing on the shoulders of giants!

In Cambridge, I boldly showed two viewgraphs entitled 'Giant Theorists', listing a selected subset of some 60 people who have investigated the structure and/or evolution of giant stars (often, many times) in the last 64 years, starting with Öpik (1938). (I was just one year old at that latter time.) My listing was produced in partial response to Douglas Gough's request for 'a balanced presentation', of which more later. The earlier parts of the listing were dominated by those who actually produced full stellar models of various degrees of complexity. These models were made initially by hand or by mechanically assisted calculation, and ultimately by automatic electronic computer, as pioneered by Fred himself. After that really hard, early pioneering work was done, the field essentially split in two. One side consisted of those who did seemingly endless large-scale computer production runs of evolutionary model sequences, which added substantial (if sometimes superfluous) detail, but little or no essentially new theoretical insight. To be ruthlessly even handed, the other side consisted of theorists who themselves mused on endlessly about their own hardly convincing 'reasons' for red-giant behaviour. Most of the latter claimed to 'explain', but in fact did not explain, self-consistently *ab initio*, either the structure of red giants or why stars became red giants. Many of the 'explanations' would seem to have more to do with some mathematical quirk or property of the solutions than with the essential physics bringing about the unusual or unanticipated behaviour of red giants. One such long-lived class of 'explanations' will be considered – and rejected as incomplete and having no particular predictive value – in Appendix 10.B. I shall also address another 'explanation', which gave as the *cause* of giant expansion what I shall show to be its *consequence*.

Not far from the meeting's venue, at one of the famed Observatory Club tea meetings, Fred once started a talk by saying, 'Oh, Ooh, basically a star is a pretty simple thing.' And from the back of the room was heard the voice of R. O. Redman, saying, 'Well, Fred, you'd look pretty simple too, from ten parsecs!'

Nevertheless, I hope to show that Fred was right about red giants. Not only that: it also turns out that I have been able to extend his treatment significantly. This extension brings about a unification of the three main stages of evolution in the life of a low-mass star: the main-sequence, the red-giant, and the horizontal-branch stages, with implications, too, for the evolution of more massive stars. In order to do this, I may have to omit a few of the crucial details, but I believe that the essence will become quite clear. But first, let's look at the context for Fred's own work in stellar structure.

10.2.1   *The preliminary milestones: the context of Fred Hoyle's early work*

What were the milestones in Fred's early research on stellar structure, particularly, but not limited to, red giants? Right from the start of their personal collaboration, Fred Hoyle and Ray Lyttleton can be seen laying down the basis for much of the future work in stellar structure that Fred subsequently carried out. They were the first to calculate what one might call complete, 'closed' models of homogeneous stars (Hoyle and Lyttleton 1942a). Their models incorporated self-consistently the density and temperature dependences learned from nuclear physics as part of the physical specification. Although both the composition of stars and the energy-generating reactions are not necessarily in accord with to-day's ideas, Fred and Ray showed that both the luminosities and the radii of stars could now be calculated *ab initio* from first principles – an important conceptual advance over all previous work. In effect this paper also included for the first time the complete, 4-equation homology relations for homogeneous stars, still useful in teaching understanding of their behaviour and overall dependencies today.

Their next joint paper (Hoyle and Lyttleton 1942b) was on the nature of red-giant stars. In it they presented a limited number of models of inhomogeneous ideal-gas stars. They later put out a related 'Note on stellar structure' (Hoyle and Lyttleton 1946). In the latter they pointed out that in the famous paper by Schönberg and Chandrasekhar (1942), the matching conditions employed at composition discontinuities were wrong. So, they were quite prepared to tackle the giants of their own time as well.

At this same time came a very interesting and significant paper (Hoyle 1946), which I think has not received enough attention. This was a most prescient paper on the chemical composition of the stars. It is very important to recognize that at the time Fred did this, many people, in particular those generally interested in stellar structure, still believed that the hydrogen content of the Sun, X, was quite low – about 0.35 by mass fraction. Schwarzschild himself published a model of the Sun that same year: it had $X = 0.35$, and absolutely huge amounts of metals. *That* was the standard picture at that time. (The progression from Eddington's assumption of heavy-element domination to a mainly hydrogen-containing Sun was far from complete. In the 1930s Strömgren played a prominent part in challenging and modifying Eddington's claims. His conclusions were subsequently championed by Chandrasekhar in his influential book published just before the Second World War, in 1939. Hydrogen content $X = 0.35$ and substantial accompanying metals of the order of ten per cent or more were essentially where things stood, in the minds of most stellar theorists.)

Only somewhat aware of this history, I found myself bowled over when I looked at Fred's paper. He notes that there were growing indications in other areas – studies of both the interstellar medium and stellar atmospheres – that the previously reigning view could not be right. He then points out, in paraphrase, 'Oh, Ooh, you get a much better fit of model main sequences to the observations if instead of something like 35% hydrogen and huge amounts of metals you assume that the stars generally contain less than about 1% by mass in the form of heavy elements and a really substantial amount of hydrogen.' As one sees by looking at his figures, the fit for low metals and a value for μ (mean molecular mass) between 0.5 and 1, and indeed closer to 0.5, passes through or closer to the general run of most of the points than those of the earlier presumed compositions. So Fred was the first stellar theorist to champion what with some modification is essentially the present picture – quite a revolutionary point to be making at that time, and one in which he perhaps went to extremes. (Indeed, in the summary of his paper he stated, 'The data suggest that at the time of condensation of the stars at least 99 per cent by mass must be in the form of hydrogen.' He would employ such a figure as late as 1959; see below.)

That particular year, 1946, was only shortly before the Steady-State theory came out. In the Steady State, of course, it was philosophically satisfying and convenient to have all the newly created material be hydrogen, i.e. of the simplest conceivable atomic form. I'm inclined to think that Fred's desire to have a unified picture, together with his stellar-structure argument as just presented, is what really led him to largely favour a high hydrogen content (and indeed, a *very* high hydrogen content), in the oldest stars in particular, for quite some time. This was a point of view he continued to favour when I became his student in 1960.

In thinking about this question of hydrogen content, I'm reminded of an image that sticks in my mind after talking with my UCSC colleague Don Osterbrock about his own memories of Fred. Osterbrock recalls one particular occasion during his days as a post-doc at Princeton. He and Fred went to an afternoon tea party at the Spitzers' home in the grounds of the Princeton Observatory. There was a most impressively large child's swing in the garden. Fred got on it and began vigorously pumping himself higher and higher. He swung so far and so high on it that the ropes became slack at the extremities. Osterbrock says he began to fear for Fred's life. Recalling that image now, I think that perhaps Fred also swung a little bit too far the other way where stellar hydrogen content was concerned. Indeed, in an oft-quoted pair of later papers (Haselgrove and Hoyle 1959; Hoyle 1959) he still used and promoted additional arguments for preferring that previously mentioned hydrogen content of 99% for the oldest Population II stars. I'll return later to discuss this point.

But I am getting ahead of the stage that we had reached in the story of Fred and his work on red giants. Hoyle and Lyttleton (1949) returned to work on the structure of stars of non-uniform composition. They published a very extensive survey paper on such models. This is the main paper in which they showed that they could obtain much more extreme sizes for stars by having various amounts of low-μ material, on top of a higher-μ and generally much more massive deep interior. (This picture was perhaps developed with a view to having the outermost, low-μ regions be accreted interstellar material. In contrast, of course, modern giant models generally have cores of lower mass with μ-values made higher by nuclear processing, and more massive envelopes that still retain their initial low-μ composition.) Hoyle and Lyttleton's figure showed the location of some of their main-sequence models in the theoretical HR diagram, and, in contrast, the positions of the much larger models they were able to obtain in their scheme.

They also produced a simple physical argument for the much larger radii they obtained from their models of non-uniform composition. They pointed out that if one puts low-μ material on the outside of a given star, that is equivalent to taking a star of a given uniform (higher-μ) composition and somewhat larger mass, and replacing the additional outer mass of *that* augmented model by material of lower μ. The resulting model will have a larger radius than the original augmented one for the following reason. Temperature and pressure are necessarily continuous at an interface. Imagine, in the first instance, maintaining these at their original values when the replacement occurs. For ideal gases, when you replace the outer layers (of original high-μ composition) by lower-μ material, the local gas energy per unit mass (inversely proportional to μ) is necessarily greater than it was before. Thus, by the continuity conditions, it has more internal energy, relative to the local gravitational energy per unit mass, and is therefore in a more favourable situation energetically to 'escape' from the rest of the star. So the star becomes bigger as a result. Carrying that mean-molecular-mass-change argument along, while not the whole story, is certainly quite a substantial component of the reason for stars becoming red giants.

As I indicated above, Hoyle and Lyttleton (1949) plotted their results in a theoretical HR diagram. The adjacent paper in the journal, funnily enough, is also on red-giant models with chemical inhomogeneities (Li Hen and Schwarzschild 1949). However, since one coauthor was Schwarzschild, I'm sure you can all anticipate the diagram that he included, the only diagram in their paper. Yes, it's the diagram that most of us in my generation grew to know and perhaps detest rather than love, the then computationally useful homology-variable or UV plane. I think these two diagrams nicely illustrate the main emphases that these future collaborators had at that time. Fred was primarily very, very interested in finding models that would really match the observational data, thereby allowing other questions to be addressed, like the ages of the stars and thus of the Galaxy. Schwarzschild, on the other hand, was more concerned with the process of calculating good models, using what, in his referee's report to my 1966 horizontal-branch paper, he called 'that gruesome tool, the UV plane'!

The preceding studies wrung what changes they could out of static discontinuous models with an ideal-gas equation of state. A few years later Sandage and Schwarzschild (1952) evolved more realistic models that were exhausting central hydrogen and leaving the main sequence. After a very promising start, those models, which had fully radiative exteriors, just rushed off to very large radii with ultimately peaking and then *decreasing* luminosities.

Two main reasons for some of the problems with these models were pointed out later by Hoyle and Schwarzschild (1955). Because of the lack of a realistic outer boundary condition (the one that would ultimately be a key feature of that later paper), evolutionary developments in the envelope proceeded too rapidly. Also, electron conduction at incipient core degeneracy was ignored. Consequently gravitational energy was released too quickly, maintaining an appreciable temperature gradient in the condensing core and thus affecting the models in a non-negligible way. Hoyle and Schwarzschild's models would correct this, making it clearer that essentially iosothermal, degenerate cores would in fact be produced.

Despite these problems, one particularly significant figure in the paper by Sandage and Schwarzschild (1952; Fig. 2, p. 471) does illustrate the generic kind of behaviour with time that occurs in the deep interiors of such stars. It shows what was in essence going to become the main puzzle of red-giant development: namely, when the inner parts or mass shells of their models moved inwards, the outer parts moved outwards. Their diagram shows a fairly clear division between the two sets of inner and outer mass shells, behaving in these opposite ways. Thus, when one asks the classic old question: 'Why do stars become red giants?' there is an alternative way of phrasing the problem. This, I think, is the real essence of the puzzle being formulated: 'Why on Earth is it that when the core of a star that has lost its previous energy sources contracts (itself intuitively reasonable), the outside perversely expands, and not only that, but expands so vigorously?' This behaviour clearly seems to involve a *bifurcation* of some sort, an observation to which I shall return.

Another figure shows the paths of their models in a theoretical HR diagram. The coolest models are still accelerating their evolution to somewhat lower luminosities and much lower surface temperatures as the latter pass through 1600 degrees. They are moving very rapidly to extremely large radii.

(In 1962 I had the great experience of attending a very fine Enrico Fermi summer school on stellar evolution, in Varenna sul Lago di Como. There, in unpublished remarks, Stefan Temesváry revealed that he and Ludwig Biermann had also performed a calculation like this, finding similarly that the model stars started leaping to the right in the HR diagram, towards very large radii. I can still recall the dramatic sweeping gesture he used to describe the models' behaviour. The two of them decided not to publish what they had done, because they were so embarrassed that it obviously failed to match what really seemed to happen with the observed giants. So, all hail to Sandage and Schwarzschild for publishing what they did, even if it ultimately 'didn't fit' the observations.)

10.2.2   Hoyle and Schwarzschild: the great collaboration

The stage was now set for the great collaboration between Fred Hoyle and Martin Schwarzschild. But I also have another subsidiary theme, and it is one hinted at by Wal Sargent in answering a point after his talk in Cambridge. Our subject abounds in serendipity – astrophysicists often end up doing great things after starting from what may be a somewhat misplaced initial motivation. In Fred's case, he has described the motivation for the work he did with Schwarzschild, in his fascinating autobiography (Hoyle 1994; pp. 282–3). He describes himself puzzling and wondering about a certain technical point, as he drove across the USA from Death Valley to Princeton. Milne and Eddington had had a famous dispute about the significance of the outer boundary conditions for stars in determining their final properties. As far as main-sequence stars were concerned, Fred and Ray Lyttleton had long confirmed the essence of Eddington's insight. But Fred was now wondering whether Milne wouldn't finally prove to have been right, at least for this very different case of the red giants. He writes,

Could it be the case, I now wondered, as I thought about the models obtained by Sandage and Schwarzschild, very far removed from main-sequence stars, that for these Milne was right?

Fred admits having had some confusion about Milne's main thrust a few pages later (p. 285). He writes,

Only once do I recall Martin Schwarzschild and me having even a moderately sharp word. This was when I was injudicious enough to say that I thought it curious that we had hit on a situation in which E. A. Milne was right in his argument with Eddington. No, Martin said, Eddington had been concerned with calculating stellar luminosities, and these were quite unaffected by the surface boundary condition. It was the radii of stars that were affected, and these had not really been Eddington's concern.

I recently had an interesting email correspondence with Leon Mestel on this topic. He tells me that he and Bernard Pagel have some recollections about Fred's discussions of these matters in the late 1950s. The upshot from their recollections seems to be that as late as about 1958 Fred was still claiming in lectures and conversation that the red giants were a case where Milne had been right. I find this a bit puzzling, personally, because by the time I began studying stellar structure with Fred in 1960 he certainly didn't seem to espouse this viewpoint. He was simply adamant that the luminosities of stars were of course determined by what was going on throughout, or deep inside them, and that, in effect, it was only the outer radius that was affected by the surface boundary condition. This is essentially a paraphrase of what he and Schwarzschild had written in their seminal red-giant paper, in two specially displayed sentences on their p. 20. (What could be claimed to be an exception to this – the radius-parametrized luminosities of stars contracting down the Hayashi track – still lay in the future then, of course.)

When Fred arrived at Princeton in 1953, Don Osterbrock was there as a post-doc after obtaining his Ph.D. with Chandrasekhar in Chicago. (Despite Fred's confident assertion in his autobiography, Osterbrock was never one of Schwarzschild's students.) He had just solved the problem of the internal structure of red-*dwarf* stars (Osterbrock 1953), and indeed had shown the importance of having the correct photospheric outer boundary condition for such cool stars. This would, of course, have fascinated Fred. The motivation that Osterbrock outlined for his investigation at the start of his paper seems a little strange to me, when read today. (But that may reflect what I have absorbed from Fred.) Nevertheless, his approach was sufficiently broad and flexible that he *did* arrive at the solution to the problem, long a standard and indeed celebrated part of the 'low-main-sequence red-dwarf' paradigm. The effects he obtained were ultimately traceable to the consequences of the sharply dropping surface opacity (largely due to the H^- ion) on the physical, radiative surface boundary condition – essentially $\kappa p \simeq g$ – as effective temperatures become ever lower. As the opacity rapidly decreases, the pressure and therefore the density at the surface sharply rise, inducing fairly deep outer convective zones. Until Don's work, makers of full stellar models had almost invariably used the simpler (and sometimes misleading) boundary condition that pressure and temperature tend to zero together. While good enough to represent the main effects for the bulk of stars with radiative envelopes, it simply fails to give a good account of those with convective envelopes. Radiative envelopes don't need a radiative boundary condition; for convective envelopes it is vital, providing a key normalization they otherwise lack. Even today, this is not always appreciated.

In Don's opening paragraph he remarks that 'models incorporating [the pp reaction] and computed under the usual assumption of radiative equilibrium always give, *with any reasonable composition* [my emphasis], a computed luminosity much higher than the observed luminosity.' Indeed, when he looked at the possibility that these stars might have very extensive and perhaps even deep outer convection zones, it was initially from the point of view of seeing whether this structural difference *alone* would reduce the central temperature, and hence luminosity, sufficiently, as Strömgren and Biermann had suggested. However, at least some of the models referenced as having 'reasonable compositions' still employed the older, much lower pre-Hoyle (1946) values for hydrogen content, even though they had been calculated in the 1950s. It is my belief, checked by homology relations, that for those particular models the compositional differences were largely responsible for the previously excessive luminosities. (Previous luminosity excesses in the range of ~12–15 can be explained this way.) It would take a change in the effective polytropic index over most of the star to effect such a result without compositional changes. That would seem to involve a much greater global structural change than the outer convective envelopes, fairly deep though those might be, ultimately found for stars of order a solar mass or somewhat less.

Some work I did in my own thesis computations a decade or so later (Faulkner 1964) relates to this point. I investigated the changes that would occur in models in which conditions in the outer parts of main-sequence stars were altered by extremely large factors. These changes simulated large uncertainties in the treatment of outer convective zones in stars of order a solar mass or somewhat less. Despite very substantial changes in the (p, T) relationship in their outer regions, the central temperatures, pressures and luminosities of these low-mass main-sequence stars were hardly changed. The main effect was solely on the outer radii, and through that the effective temperatures. More extensive outer convective envelopes, all else being unchanged, led to significant reductions in overall radii, relative to their radiative counterparts, as the thickness of such pseudo-convective regions understandably decreased, relative to their radiative counterparts. So the general conclusion from my main-sequence work was that the models mainly shifted to the left in the HR diagram as more convection occurred in their outer layers. Later work by others, increasing the efficacy of convection by increasing the value of the mixing-length parameter, led to similar results, whether on the main sequence or the red-giant branch (RGB).

This was of course re-obtaining the point – the reduction in outer radii brought about by convective external regions – that Fred himself had seized upon from Osterbrock's work, the key thing about it that he would emphasize in later years. (Curiously, Osterbrock himself did not explicitly make the point about reduced radii in his own paper, although it is there implicitly, of course. He concentrated mainly on resolving the models' luminosity discrepancy rather than their radius mismatch. I do not recall Fred ever saying 'Osterbrock solved the problem of the red dwarfs' luminosities' – his remarks were always about the mismatch in their radii, and that is how he would consistently describe Don's paper and his insight.) However, it also isn't clear that Fred himself saw the implications as quickly as might appear from reading his autobiographical account.

Both Schwarzschild and Osterbrock have reminisced with me about those heady days. Schwarzschild once described how Fred would come into his office at the Princeton Observatory early in the day, brimming with new ideas or a new twist on an old idea to be tried in the attempt to understand red giants. Schwarzschild and his long-time hand-calculational collaborator Richard Härm would then calculate away madly, sometimes showing by the end of the same day that what had looked like a good suggestion wouldn't work.[2] Never daunted, Fred would be back again the next day, and the procedure would be repeated. Schwarzschild was very impressed by Fred's resilience and continued inventiveness. Eventually, according to Schwarzschild, in came Fred with another idea that he and Härm couldn't shoot down, and that became the key to their ultimate success.

[2] Note that all this heroic work, including that carried out by Hoyle and Schwarzschild together, was done in the days *before* the automatic computation of stellar evolution, a later procedure first introduced by Haselgrove and Hoyle (1956a). Whether aided by mechanical hand-cranked or electric calculating machine, there were very many tedious steps to be taken by hand. So hats off to all of them!

Osterbrock in turn recalls many conversations with Fred. Their rather ancient offices were next to one another in the Observatory, somewhat removed from Schwarzschild's. Fred would often pop in to talk in Osterbrock's office; Osterbrock would do the reverse, though somewhat less often. On many of these occasions, Fred would ask him question after question about his work on the red dwarfs. Osterbrock has ruefully said he was flattered that the great man should want to know so much about his work. He, Osterbrock, had no idea that Fred was going to apply his approach to the problem of understanding red giants, and he has said later that he could have kicked himself that this application didn't occur to him. Osterbrock was aware that Fred and Schwarzschild were engaged on some great red-giant enterprise which involved many long discussions, but he doesn't seem to have known that the implications of his work on red dwarfs were to play a significant part in the success resulting from their collaboration.

In Hoyle and Schwarzschild's ultimate work, the effects on the positions in the HR diagram of later, evolving post-main-sequence stars were most dramatic. If radiative envelopes were computed according to the former, unphysical boundary condition, they expanded out to grossly huge radii such that nothing like the correct surface boundary condition could ever be met. Densities for any putative effective temperature were always far too low. In contrast, the proper surface boundary condition, by firmly relating conditions in the photosphere and giving much higher densities there, induced convection in the outer regions and ultimately extremely deep outer convection zones. Their completed models climbed up the cool regions of the HR diagram, all the way to the previously unreachable tip of the red-giant branch – a triumphant and seminal result.

Although it is the surface boundary condition that induces these effects, it is also worth looking at what is happening in the star from the opposite point of view: from the inside out, as Hoyle and Schwarzschild sometimes did in their paper. Deep inside the star, starting just outside the burning shell, is a region that is extensive in radius yet ultimately containing very little mass. (I shall describe the way this feature impressed itself upon Fred in the next subsection; it was in fact encountered in their computations 'from the outside in'.) Passing further outwards, the previously radiative envelope becomes convective at some point. Beyond this point, densities first drop much less steeply than they would have done in their radiative counterparts, i.e. as $\rho \propto T^{1.5}$ rather than $\rho \propto T^3$ or so. Thus densities at successively lower temperatures are now increasingly larger than in their radiative counterparts, leading to far more efficient mass storage. On the other hand, at still lower temperatures another factor comes into play: extensive helium and hydrogen ionization zones in which the corresponding adiabatic (or near-adiabatic) slopes for ρ vs. T are much steepened – the consequence of what is more often referred to as the lowering of the adiabatic temperature gradient. There, densities drop much faster with temperature (though from a higher base) than in the previously conventional radiative treatments. Thus, hitting the correct physical surface boundary condition 'from the inside out' becomes a tricky and rather circuitous business. To illustrate this in their paper, Hoyle and Schwarzschild considered the two extremes of having either completely radiative or completely convective envelopes (by which they meant log ρ − log T traces with the appropriate slopes), all the way from what they knew to be the density and temperature conditions deep down near the core, out to a putative surface. They showed that such extreme bracketing cases resulted in densities at T_{eff} far too low or far too high, respectively, to match the photospheric boundary condition. Hence they concluded that a red giant could 'adjust its photospheric density to the required value by adopting a structure with a sizable outer convective zone, the parameter of adjustment being the depth of this convective zone'. Note that the emphasis here was on obtaining a correct, if devious, trace in the log ρ − log T plot, whether ending or starting with the right photospheric conditions. There was no reference to mass storage as a major factor or consideration it per se.

In the 1960s, Fred would tell his students that at the time of his work with Schwarzschild he saw the problem mainly as one of properly satisfying the surface boundary condition, and understanding its consequences for the layers beneath it. That, he would say, was an 'outside-in' way of thinking about it. (In their paper, he and Schwarzschild had switched back and forth between their outside-in or inside-out arguments, making it a little difficult to decide which viewpoint, if any, had primacy in their minds. Perhaps they even differed in this. Nevertheless, in the Schwarzschild–Hoyle 'shooting method' for computing models in those pre-Henyey-method days[3] it was natural to continue integrations downwards from a satisfied boundary condition at a putative surface for 'matching' in the deep interior.) Later, however, Fred came to view the red-giant structural conundrum as more one of understanding how the envelope mass was to be stored, given the conditions holding deep down near the developing

---

[3] The latter should perhaps more properly be called (by astrophysicists) the Henyey–Wilets method, if names are to be attached to it, but I suppose it is far too late to change that. Henyey had asked his former Manhattan Project colleague and friend Wilets if there wasn't a way that distributed knowledge of a reasonable prior approximation couldn't be exploited to obtain a better approximation to a new solution under slightly changed conditions. Wilets thought about it, and came up with the approach now known to astrophysicists as the Henyey method (Henyey et al. 1959). The noted English engineering scientist Southwell had long used 'relaxation methods' (possibly inventing that term) for complex structural problems or those like the bending of elastic plates under various loads (Southwell 1940), as I myself heard in Cambridge Tripos Part III lectures in 1960. However, that knowledge had clearly not filtered through to the astrophysical community.

core – an 'inside-out' problem. Naturally, both of these problems or conditions must be met in any completed model, but it was ultimately consideration of the red-giant problem from the latter, envelope mass-storage point of view that led him to the approach studied and enlarged upon later in this paper.

Having now dealt at some length with the main conceptual breakthrough in Hoyle and Schwarzschild's paper, let me just mention one or two oddities or lacunae, some of which led on to future work by Fred and others. They made various approximations, some intended and justified, and one perhaps not intended. For computational convenience, they assumed that the shell temperature was constant at 20×10^6 K for a substantial part of the red-giant branch, up to the level of about the horizontal branch (HB). Beyond that point, they allowed it to vary and found that it essentially climbed in lock-step with the core-mass. (This feature is found in my theory for most of the RGB.) There's a puzzle about the value of X_{CN} (their terminology) used in their Population II models. Unless there is simply a misprint, two different values were used (0.0005, p. 10; 0.005, p. 19) when considering the earlier and later parts of tracks for the same Population II models. A further oddity occurs when they examined the effects of increasing the heavy-element content to that of Population I stars. They certainly incorporated that into their envelopes, deducing the way in which it would affect surface characteristics. However, read literally, they appear from their description to have taken certain deep interior information from a previously presented table, when computing their 'complete Population I' models. That table contained data from their earlier interiors computed with the Population II metal content (p. 30), although its actual value is subject to the uncertainty already mentioned. So it is possible that models for one of the computed sections, or perhaps a whole evolutionary track, combine together inconsistently computed surfaces and interiors.

In a more speculative ending to their paper, Hoyle and Schwarzschild explored the properties of what they took to be post-He-flash models as possible models for HB stars. They found themselves in something of a paradoxical situation. In part because of using $X = 0.90$ (but more for reasons of nuclear physics – see below), even their 'metal-weak' models with core-masses as high as $0.72\,M_\odot$ still remained firmly on the RGB, although they did slide down to about the luminosity level of the HB. (It must be noted that the identifier 0.60 shown in their Figure 5 is not the core-mass, but rather the core-mass *fraction* for a star of $1.2\,M_\odot$.) Not until their core-masses were of order $0.80\,M_\odot$ did their models leave the RGB; they were then quite a bit too luminous to be HB stars. There was also, as they pointed out, a problem with the available timescale there. The conversion of helium to carbon yields only about one-tenth the energy per gram as the conversion of hydrogen into helium. Therefore, all the central helium would have had to be consumed in their models long before they'd had a chance to detach themselves from the RGB. Thus, although Hoyle and Schwarzschild's figure shows that quasi-static models with very large core-masses could occupy more luminous parts of the Hertzsprung gap, it appeared that this was *not* the way to produce models for the observed horizontal-branch stars.

This was an apparent problem, or paradox, which I was able to resolve in my first paper on the horizontal branch (Faulkner 1966). I realized that since the time of Hoyle and Schwarzschild's original paper, a significant change had occurred in the understanding of the details of the CN cycle. Consequently, the energy generation rate previously attributed by them to the CN cycle had been too large by a factor of $\sim 10^2$. (Physically, it had been assumed that $^{14}N(p, \gamma)^{15}O$ was a resonant reaction; by 1960 it was known that it is not, and that it, as the slowest part of the cycle, made the rate much smaller than before.) When I made my own HB models with the new, now reduced CN energy generation rate, they could detach themselves very nicely from the RGB for low Population II metal contents, but for Population I ($Z \simeq 0.01$) most of them were still jammed up against the RGB unless their envelope masses were extremely small. With the revised rate for the CN cycle, even the lowest values of Z employed by Hoyle and Schwarzschild using the old rate would correspond to energy generation rates for the highest Z-values, or even much larger amounts, according to my own computations. No wonder, then, that they obtained their two major mismatches. As viewed with this hindsight, their models had been jammed onto the RGB for far too long because of their extremely high CN-cycle energy generation rates for a given X_{CN}; and when they did get off (as statically computed survey models, not evolutionarily connected sequences) they had both excessive luminosities and a timescale problem for getting to such configurations in practice. (No wonder Fred said, 'You're tackling quite a challenging problem', when I told him once I had graduated in 1964 of my intention to look into this still dangling aspect of their work.)

After his great collaboration with Schwarzschild, Fred was to revisit this post-He-flash problem again with Brian Haselgrove (Haselgrove and Hoyle 1958). However, in those calculations, which extended their own work on the prior stages (Haselgrove and Hoyle 1956b) by just another seven time-steps (in two years!), they still employed the older value for the CN-cycle energy generation rate, with $X = 0.93$ and $X_{CN} = 0.0025$, energetically the equivalent by my time of using something like the huge and physically impossible value 0.25 for the latter, in round terms. Consequently, it was not surprising that their 'post-flash models' *still* adhered rigidly to the RGB. (This was very unfortunate. In another paper to be discussed briefly (Hoyle 1959), Fred attributes the improved understanding of the CN rate to B^{2}FH – Burbidge *et al.* (1957). But that change does not seem to have caught up with his stellar-structure computations until 1959, after the work reported in Haselgrove and Hoyle (1958).)

Haselgrove and Hoyle (1956b) had taken the early stages of their first computations to the point of yielding an age of 6.5×10^9 years for the globular cluster M3, a result little different from the value of $\sim 6.2 \times 10^9$ years that he and Schwarzschild had generally obtained for Population II stars with hand computations. (I should note that the immediately preceding paper by Haselgrove and Hoyle (1956a) contained a description of the very first automatic computer programming method for following stellar evolution.) But with the revision in the CN energy generation rate, Haselgrove and Hoyle (1959) laid the foundation for an oft-quoted result in Fred's immediately following paper (Hoyle 1959). In that, from his computations and a rather involved argument invoking the brightness of RR Lyrae stars, he first announced to the world an italicized result that he knew would be considered startling at the time: 'The age of the Galaxy is in excess of 10^{10} years'. (His formal result implied an age $t > 1.2 \times 10^{10}$ years.) Fred's paper has often been quoted for this eye-catching result, but no-one who does so ever seems to note that such a large value for the age of the Galaxy was obtained only for parameters that would raise eyebrows today: $X = 0.99$, $Y = 0.009$, and currently evolving masses quite a bit above a solar mass ($M > 1.3\,M_\odot$). For $X = 0.75$, much closer to today's understanding of the hydrogen content of Population II stars, he obtained only $t \simeq 4.8 \times 10^9$ years, a value his additional RR Lyrae argument led him to reject. I have always felt it was a little disingenuous for later authors to quote the much higher age without even the tiniest caveat paying attention to Fred's choice of parameters.

In the end, however, Fred was right about the age of the Galaxy, but not for the range of parameters or the argument he had used. The realization that horizontal-branch systematics argued against very low helium content lay in the future (Faulkner 1966; Faulkner and Iben 1966). I had already formulated a different argument for the low values of currently evolving masses found in the latter paper, as a necessary concomitant of the high helium content preferred in the former paper. That simple argument was not included in either of these papers, but would appear subsequently in a form making it look like an afterthought in Faulkner (1967). The latter obscurely titled paper on the 'Eggen–Sandage residue' showed that there was indeed previously unappreciated evidence even in main-sequence systematics for 'high helium' in all main-sequence stars including the so-called subdwarfs. That paper also introduced a new parameter, the quantity $\Delta Z / \Delta Y$, broadly describing overall enrichment in later generations of stars from some initial (cosmological?) base of $Y = Y_0$, $Z \simeq 0$. It would become a focus of much later research, investigation, and determination. I found its value then to be of order 3. Stan Woosley, one of the earliest young visitors to Fred's

Institute and a UCSC colleague today, has said that this is still the best overall value for it (though there is some spread, of course) in both observations and what emerges from supernovae. The end result of combining significantly lower masses ($M < 0.8\,M_\odot$) with high helium content for old, currently evolving stars indeed yielded ages for Population II stars in excess of 10^{10} years, as Fred had first found for much different parameters and from using a different argument.

10.2.3   A balanced presentation

When inviting me to give a paper, Douglas Gough asked that I should give 'a balanced presentation'. I believe this meant that I should give some weight to the variety of explanations proposed in answer to the classic question: 'Why do stars become red giants?' The trouble with this is that scores of people have written on red giants, and their own presumed reasons why stars become red giants. Thus, a truly balanced presentation would require far more space than I have at my disposal. So really, on looking at all these theoretical papers, the only way I could come up with a balanced presentation (or at least, two points out on opposite ends of a logical seesaw, an image I shall employ to other and greater effect, later) was to quote, in summarizing paraphrase, from two of them. At the very end of their paper on this topic, the conclusions of Refsdal and Weigert (1970) are essentially that:

(i)   Red giants are very large because they are very luminous;

whereas Whitworth (1989) in his abstract concludes that:

(ii)   Red giants are very luminous because they are very large.

So much for balance.

Faced with such penetrating and utterly opposed theoretical insights, one is almost rendered speechless – but, given your present interlocutor, you need have no fear of that. I might also remark that Douglas Gough is hardly the person to ask for balance, since I first met him when he tripped and stepped in my soup while walking across the mathematicians' dining table at St John's College.

Consider a schematic model of a low-mass red giant, the kind of red giant that interested Fred so much. This model is indicated in Figure 10.1. It is of course sketched with the benefit of hindsight not available to Fred when he began working on this problem. However, I hope to show that from the simple starting point I give below, it is possible to derive the necessity for all the various regions into which we now realize such a star should be divided. The need for their existence, as well as their relative extents in both mass and radius, all follow from my starting point in a well-ordered logical and deductive argument.

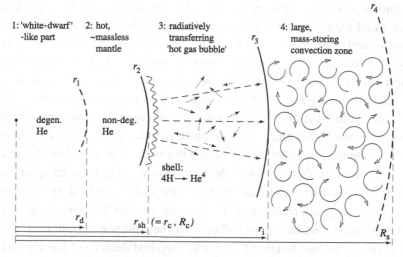

Figure 10.1 A schematic model of a low-mass red giant, showing its four distinct extensive regions. A fifth region, the shell, is taken to lie at the base of region 3 in our approximation, just above a composition discontinuity. Though idealized as having no radial extent, it essentially determines the rest of the structure.

(For those curious about this, the starting point involves assuming that a dense core develops following central hydrogen exhaustion, and analyses just what we mean by a 'dense core', in context – see the discussion surrounding equation (10.1), some pages ahead.)

The typical low-mass red giant can be divided up into at least four extensive regions, two inside the nuclear-burning 'shell', two outside. The shell itself represents a fifth, special region. As I shall show, not only is this a region rather different from the others, but, detached as it is from the centre of the star, it is largely responsible for the phenomenon of giantdom. (As I shall also show, the fact that it sits on top of a dense core is the other main factor.) One could treat the shell as having its own characteristic internal structure, as Eggleton has done, for example. However, for simplicity I shall treat it as effectively a very narrow entity, essentially situated at a composition discontinuity. Its composition will be taken to be that of the uniform envelope above it (that uniformity being another approximation, of course) and its temperature, T_{sh}, that of the very base of the envelope. Its radius, variously written as r_{sh}, r_c, or R_c, according to context, is taken to be that of both the base of the envelope, again, but also that of the complete (helium) core, which itself is divided into two rather different regions.

This simplified treatment necessarily involves certain physical discontinuities at the shell, envisaged as a mathematical interface. In the real world, such a discontinuous interface does not exist, of course. In practice, there are variations in all physical variables throughout the nuclear-burning region, and there is some ambiguity about deciding whether any one particular point inside it may be taken as 'a characteristic point' corresponding to an interface. For example, in fully computed models, the points of $0.5L_{nuc}$, $0.5X_{env}$, $\max(\epsilon_{nuc})$ etc. are all different, and indeed the nuclear-burning rate ϵ_{nuc} peaks where the local hydrogen content is already quite low, but where the temperature and density are somewhat higher than those obtaining at the 'base of the envelope', itself a place requiring another operational definition. Thus one should realize that some special care may ultimately be needed in making strict comparisons with actual models. Fortunately, however, insight may be gained by continuing in the spirit and proud tradition of those astrophysicists like Eddington, Hoyle and Schwarzschild, who have gone before with their pioneering work. In particular, there vividly springs to mind Eddington's spirited defence of the applied mathematician/physicist who leaves the tool-marks on his finished work, rather than the pure mathematician intent on producing a shining alabaster end-product with no indication of its means of production (Eddington 1926).

We turn now to the rest of the structure. Starting from the centre, we have:

(1)   A helium core out to a radius r_d dominated by degeneracy, in which therefore the value of the temperature plays no effective rôle. Almost the entire core-mass resides in this part. We may call it the 'white-dwarf-like' part of the red giant.

(2)   Outside this, but still below the shell, is an essentially isothermal ideal-gas helium region. The temperature in this 'hot helium' region is close to that of the shell, and rather little mass is stored in it. (This is analogous to the usual situation in a classical white dwarf, where only a negligible amount of mass can be stored in the outer non-degenerate regions, even though the lower parts of those regions may have temperatures in the tens of millions of degrees K, and densities of the order of a thousand g cm^{-3}.)

(3)   Further out in the red giant (and now outside the shell) is a 'radiative region', again containing very little mass in the limit as core-mass increases, but with substantial drops in temperature, density and pressure through it. Fred would call this a 'hot gas bubble', although, as my approach shows, the mass in this region can be fairly substantial for stars with low core-masses in their early stages of red-giant-branch evolution. My analysis shows explicitly that its mass drops extremely steeply with core-mass as evolution proceeds up the RGB. Consequently the convective envelope (next region) digs down to

a greatest depth in the mass variable, and then retreats back to higher values. This result stimulated my theorem of maximum penetration.

(4) Finally there is a large convective region (the 'convective attic') of very large radial extent in which the remaining mass of the star is stored. The densities in this region are extremely low in the end – although nowhere near as low as had the radiative region's dependences continued throughout it. This is the region that provides a low-mass red giant with one of its most noticeable and characteristic features: a grossly distended envelope which may make the star's radius more than two hundred (>200) times its original main-sequence size, and some ten thousand ($\simeq 10^4$) times the radius r_d of the innermost core. Large though it is, however, it is much smaller than a completely radiative envelope would be.

As evolution proceeds to late stages, both regions 2 and 3 contain very little mass. It is as though the shell has cleared mass away from either side of itself, although the details are more subtle than that simple observation suggests. By a process of finely tuned self-consistency – to be examined later – the shell determines where it will sit, in splendid isolation, controlling the structure of the rest of the star.

All this lay very much in the future as Fred approached the red-giant problem, of course. Indeed, what I have just described is the basic picture of a red giant that we owe to Fred himself. What were the essential steps leading to this picture?

Fred would talk about a repeated experience he had had when he was first computing models of an evolving, roughly solar-mass red giant. There was something about the successive attempts to find a solution that he found quite remarkable. (There was time to *watch* the solution being found, in those days, which I consider to be an advantage lost with today's incredibly high-speed machines.) Fred noticed that the calculations 'ran into a brick wall' in a certain deep region, on going inwards into the star. The pressure would go up by four, five, or even six orders of magnitude, in what I have called region 3, while the mass variable itself would hardly change. (He was very, very struck by this initially strange but ubiquitous behaviour, which became evermore pronounced as models evolved further up the RGB.) He became convinced that the only way to understand the structure of a red giant would be to appreciate, *from the inside out*, what this 'brick wall' meant, and why this peculiar structure was not only necessary but was somehow forced upon the star. And so he ultimately came up with the view that the typical red giant should be thought of as a white dwarf struggling to store a very large amount of mass in its exterior. Now everyone knows you cannot store much matter outside, or at least in close proximity to, a white dwarf. If the densities in the shell region and just outside it are low compared with the mean interior density – as in a typical white dwarf – you just cannot store a sensible amount of mass there.

In my terms, then, a red giant is basically a white dwarf with a storage problem. And like many of us with a storage problem – you will appreciate this particularly well if you have seen my house or office – there is only one thing to do: you remodel the attic. And that is what the star does – it remodels the attic.

10.3   Standing on the shoulders of dwarfs; the remarkable, bifurcated structure of stars with 'dense cores'

We are now ready to approach red giants from Fred Hoyle's viewpoint, one which enables their structure and evolutionary behaviour to be deduced *ab initio* almost as inevitable. Some may disagree; indeed, when I gave a talk about this while on sabbatical leave in Munich, Alvio Renzini said: 'Your so-called explanation of red-giant behaviour relies on the use of secret arcane methods.' Well, I learned almost all these methods from Fred, so let me show you them now. I shall lay out the basic principles behind the explanation, the algebraic consequences which follow almost immediately, and discuss some further implications. The reader can then judge just how arcane these methods are.

10.3.1   *The zeroth principle of stellar structure*

First I want to introduce you to a very important zeroth principle of stellar structure, which does not seem to have been previously recognized. Stars have *two* important centres. One is

(i)   the *geometrical* centre, of highest density;

and the other,

(ii)   the *nuclear* centre, of highest ϵ_{nuc}.

Centres (i) and (ii) are not *necessarily* the same, and exploiting this unifies the MS (main-sequence) and the post-MS phases of low-mass stars. Not only does it help us to appreciate the deep relationships between the structures of MS, RGB, and HB stars, it also has implications for the evolution of more massive stars.

The main trouble is that our 'intuition' about stars has been developed in a situation in which these two centres are in the same position. To make a short cut to the main point, the problem we have with understanding why stars become red giants is the difficulty we have in anticipating the need for, and when, these two centres will bifurcate. It is an absolutely classical bifurcation problem.

At the meeting honouring Fred Hoyle in Cambridge I announced that I had, as it happened, a simple mechanical model of a red giant right there, in my pocket. Actually, it was of course an analogue model, but it illustrates the point quite well. The model, an ingenious child's toy, was given to me many years ago by Martin Gaskell, when he was suffering through my account of bifurcation theory in a graduate classical mechanics course at UCSC. I realized its analogue applicability to the red-giant problem only when preparing the talk. The model consists of four curved metal 'leaves' hooked at their lower ends onto a short rubber band. In the resting position the leaves are closed, making an apparently innocent looking Christmas tree surrounding a diminutive and hidden Santa Claus. (Although red, not white, this usually corpulent figure is nevertheless the equivalent of our dwarf.) An attached toothed wheel arrangement below the device allows one to impart substantial spin (and angular momentum) to the leaves.

When the leaves rotate slowly, they maintain the same closed-up structure. But if you spin the connected leaves quickly, the elastic band is unable to keep holding them in, and the leaves spring out, revealing the dwarf. It is difficult to see this happen when it is being started up. It is either closed or open, the change occurring very, very quickly. However, if it is started up fast, so that it is in the open position, friction will gradually slow it down. And all of a sudden – snap! – the leaves return to the closed position. It is clear that there is a very critical point at which the leaves snap back in. The reverse of that quick collapse is a sudden expansion, and that is the analogue of what the red giant is doing, particularly if one clothes it in a purely radiative envelope.

In the mechanical example, the governing parameter is the angular momentum. The sudden switch to a newly possible series of open equilibrium configurations (while meanwhile, the old series becomes unstable) involves an orthogonal transition in an appropriate kind of phase-plane. (Ed Spiegel tells me that the term often used to describe this transition – the 'exchange of stabilities' – is a mistaken use of what Poincaré meant by this expression.)

In the stellar case, the parameter is in a sense the answer to the question: 'Where is the nuclear burning taking place?' In main-sequence stars, the nuclear burning is concentrated at the centre. The stellar structure is 'closed up'. But when the 'nuclear centre' gets far enough away from the geometrical centre, a transition takes place to a new kind of structure involving a non-burning core, a thin shell, and, *consequently*, a *very distended* envelope. (I am reminded of my dear daughter Sarah, and what she said as she tried to use an inflatable ring in the swimming pool that she had not played with for a year or two. It was very tight, and she asked the world, in great surprise: 'Who blew me up?'!) I intend now to explain why this happens to the giants. I shall then return to characterize what happens in yet one more way, after pursuing Fred's lead in what follows.

10.3.2   Some evolutionary preliminaries

Initially I shall mainly treat hydrogen-rich low-mass stars. (With appropriate changes the ideas can be applied to other stars, e.g. helium stars.) I shall assume that we can take the following set of events as given in a general way.

A low-mass star burns out its central hydrogen-rich region over a long time. As it does so, two natural consequences occur: (i) nuclear burning is increasingly concentrated in a thin shell surrounding a compositionally differentiated core; (ii) there is a tendency for that core to contract and become denser. The latter tendency can be demonstrated in a variety of more-or-less plausible ways, but it is the tendency of the envelope to expand concomitantly that is the real puzzle. Another more colloquial way of putting the age-old question: 'Why do stars become red giants?' is 'Why do envelopes become fluffier as cores grow denser?' I shall demonstrate that that *is* what happens in well established red giants.

Contextual meaning of 'a dense core'

What is meant by 'a dense core?' There is only one useful definition that makes contextual sense and enables us to make progress. Thus, by 'dense core', I mean one in which the *compactness* $\rho_{sh}/\bar{\rho}_c$, namely the ratio of nuclear-shell and mean-interior core densities, is very much less than unity. This is so important to remember that I now set it in the form of a displayed inequality:

$$A \text{ core is a 'dense core' if } \rho_{sh}/\bar{\rho}_c \ll 1. \tag{10.1}$$

In practice, while central degeneracy is a way of ensuring that this condition holds, it is not absolutely required, although that does represent the extreme case. Both moderate-mass and massive stars leaving the main sequence, or low-mass stars in the post-RG, HB (post-red-giant, horizontal-branch) phase of evolution also have 'dense cores' in the sense of equation (10.1), although none of them is centrally degenerate. Such stars may also have substantial core luminosities. My treatment has been extended to take explicit account of the effects of that on the changed resulting values for $\rho_{sh}/\bar{\rho}_c$ in comparison to those for models having zero-luminosity cores. I shall in fact show that the behaviour of horizontal-branch stars may be viewed as a 'missing link' to that of the former kinds of star, when looked at in a certain way.

Thus, our discussion of the consequences of central compactness will be extremely broad, since it will apply to essentially *all* post-main-sequence cases. In order to simplify the exposition, I first concentrate entirely on the most extreme case: the low-mass red-giant example with a degenerate core and no central-core luminosity. Only after dealing exhaustively with that example shall I broaden the discussion out to more general cases.

In all cases the value of $\rho_{\mathrm{sh}}/\bar{\rho}_c$ will be found self-consistently as part of the method; in the extreme case of the low-mass red-giant example with which we now continue, $\rho_{sh}/\bar{\rho}_c$ will have its smallest value for a given core-mass. It becomes rapidly smaller as both $\bar{\rho}_c$ and M_c become larger. Therefore, we start exploring the changing structure of red giants once $\rho_{\mathrm{sh}}/\bar{\rho}_c$ is 'small enough'. (Not only is $\rho_{\mathrm{sh}}/\bar{\rho}_c$ the *only* sensible parameter, in context, to characterize the presence of a dense core, it also turned out to be crucial in my discovery of the theorems and other results that follow.) Additionally, as far as radiative transport is concerned, electron scattering comes to dominate the opacity in a natural way. I shall assume also that we are in that limit already.

10.3.3   *Arcane principles and consequences*

We have already seen, above, the zeroth principle of stellar structure, i.e. the notion that stars may possess at least two physically important centres.

The fact that the nuclear centre lies on the surface of the core is ultimately responsible for the most exquisite kind of feedback in red giants. That very strong feedback produces the effects I shall describe. Whenever one has such a finely tuned situation, it becomes a matter of taste where one enters the circle of logic. I do so with the initial assumption discussed above, namely that $\rho_{\mathrm{sh}}/\bar{\rho}_c \ll 1$, and then I indicate how that can be checked for self-consistency.

First principle: mass is conserved

By this I mean that the star started out in life with much more mass than is now present in its core, and unless it has found some way to slough this outer mass off, it must still be stored. But it is extremely difficult to store any substantial mass immediately outside a thin shell where all the luminosity is produced, and where $\rho_{\mathrm{sh}}/\bar{\rho}_c \ll 1$. These are the very conditions that, at the outside of an 'ordinary star', call for the typical, virtually massless envelope or 'radiative-zero solution' (Schwarzschild 1958). I shall call the latter a 'classical envelope'.

Thus the conditions outside the core appear to be those for a classical envelope. Yet that envelope is charged with a most curious task: to store the embarrassingly large amount of mass with which Nature has endowed it. The key to its solution lies in the value of the only adjustable parameter left in the classical envelope's temperature–radius relationship. This is the constant τ, defined by its appearance in equation (10.2) below. In classical envelopes, τ is generally written as a *negative* multiple of $1/R_s$; in this form it is responsible for the commonly seen factor $(1/r - 1/R_s)$. Classically, even the starting value of r/R_s is 'large', i.e. of order 0.7 or more, so that τ is both negative and 'large'.

In sharp contrast, it is easy to see by a process of elimination that in the present case τ must be both *positive* and 'small', yet necessarily larger than a very closely determined minimum value. The degree of 'smallness', τ/T_{sh}, is directly related to $\rho_{sh}/\bar{\rho}_c$. Specifically, for the electron-scattering case we find

$$T = \frac{1}{4}\frac{G\mu\beta M_c}{\Re}\frac{1}{r} + \tau, \tag{10.2}$$

where,

$$(\tau/T_{sh}) \geq (\tau/T_{sh})_{crit} \simeq 3(\rho_{sh}/\bar{\rho}_c). \tag{10.3}$$

In general, mass storage above the shell is negligible when τ is less than this critical value, but can be extremely substantial once that value is exceeded. In fact, τ plays a significant rôle in determining several features of the red giant's structure which ultimately develops outside its shell-burning region. These include the increasing relative extent of the immediately surrounding radiative zone and the decreasing amount of mass stored there as evolution proceeds.

(The critical value of τ stated above is very easily obtained from the slightest consideration of mass-storing possibilities in the classical *UV* plane, once $\rho_{sh}/\bar{\rho}_c$ is very small.[4] If that is 'arcane', it is but one more example of the parlous state into which the study of theoretical stellar structure has fallen.)

Second principle: energy is conserved

By this I mean simply that in a quasi-static steady state, the luminous flux outside a shell surrounding a core equals the rate of energy production in regions below the point in question. Symbolically,

$$L_{rad} = L_{sh} + L_c. \tag{10.4}$$

This three-term equation, with $L_c \neq 0$, permits application of our general approach to other core/envelope stars such as horizontal-branch stars. That case will be treated later. However, for κ_e-scattering with *no* L_c (the usual red-giant approximation), the corresponding two-term equation simply yields

$$\frac{4\pi c G M_c}{\kappa_e}\frac{\frac{1}{3}a\mu\beta T^3}{\Re\rho} = \frac{4\pi\epsilon_0}{\eta+3}r_{sh}^3\rho_{sh}^2 T_{sh}^\eta, \tag{10.5}$$

---

[4] This critical value of τ for significant envelope mass storage is obtained in Appendix 10.A.2. In effect, in their work, both separately and together, Schwarzschild and Hoyle took τ to be zero, if for different reasons – see that appendix.

using a thin-shell approximation for $L_{\rm sh}$. With the idealized approximations I am making, the left-hand side holds true down to the shell itself, so that $T_{\rm sh}^3/\rho_{\rm sh}$ may be substituted for T^3/ρ in what follows.

10.3.4   *The asymptotic theory of low-mass red giants: stars with 'zero-luminosity' cores*

The complete joint solution of equations (10.2) and (10.5) requires iteration, but for many purposes this is unnecessary. One can, as a first step, ignore the small quantity τ. In that case, substitution of the dominant term for T (see equation (10.2)) in equation (10.5) determines $\rho_{\rm sh}(M_c, r_{\rm sh})$, where $r_{\rm sh}$ is also a synonym for R_c, the core radius.

The results obtained by doing this provide several intriguing and *explicit* 'asymptotic' power-law relationships between various red-giant parameters. I call this the 'asymptotic theory' of red-giant structure. Red-giant model properties must necessarily evolve towards these asymptotic relationships as $\rho_{\rm sh}/\bar\rho_c$ becomes self-consistently smaller and smaller with time (as I shall in fact show). Finer details of the approach to the ultimate asymptotic power relationships can be obtained by estimating τ and iterating perturbatively, if desired. (This enables one to obtain explicitly the character of the final, curving paths followed by several pairs of representative variables in plots like that in the ($\log \rho$, $\log T$) plane, for example, as physical asymptopia is approached. That is something simply not possible in other attacks on this problem.) For the most part, however, I shall concentrate on the asymptotic results from here onwards. I omit the tedious algebra that enables almost all the following equations or proportionalities to be written with explicit functional multipliers.

Recognizing that equation (10.5) can be written as

$$L = A/\rho_{\rm sh} = B\rho_{\rm sh}^2 \tag{10.6}$$

greatly simplifies the task of obtaining explicit expressions for both L and $\rho_{\rm sh}$ in terms of the other quantities summarized here as A and B. Both L and $\rho_{\rm sh}$ may be written as products of explicit powers of M_c, $r_{\rm sh}$, and natural numbers, physical constants, or other constitutive factors. However, additional manipulation yields two new *explicit* and significant results, or theorems. (Functions $f(\#{\rm s})$ and $g(\#{\rm s})$ below incorporate the other explicit factors.)

Two new, explicit asymptotic theorems

From the preceding results one can derive

Theorem 1: The luminosity–compactness–core-mass relationship:

$$L\left(\frac{\rho_{\rm sh}}{\bar\rho_c}\right) = f(\#{\rm s})M_c^3; \tag{10.7}$$

Theorem 2: The density-seesaw theorem (specifically for $\eta = 15$):

$$\rho_{\text{sh}} \bar{\rho}_c = \frac{g(\#s)}{\epsilon_0^{1/3} M_c^{8/3}}. \tag{10.8}$$

Theorem 1 is obtained as follows: Equations (10.6) lead most directly to expressions for L and ρ_{sh} in terms of other inputs. However, because I had concluded that the factor $\rho_{\text{sh}}/\bar{\rho}_c$ was of interest and significance for many purposes, I also obtained its own general form. On substituting a particular, explicit value for η (16, not necessarily the best) I noticed that the completely separate expressions I now had for L and $\rho_{\text{sh}}/\bar{\rho}_c$ contained striking combinations of other variables which would cancel on multiplication. The result, which I quickly verified was quite general, immediately emerged.

Alternative derivation of Theorem 1

Theorem 1 looks strikingly similar to Eddington's classical electron-scattering mass–luminosity relationship, though with several important differences: it involves only M_c (not M_*), and it includes the ratio $\rho_{\text{sh}}/\bar{\rho}_c$ as a mediating factor. (This factor makes the significant statement that as $\rho_{\text{sh}}/\bar{\rho}_c$ decreases in a core/envelope structure of given M_c (for whatever reason), L correspondingly increases, in direct inverse proportion. Thus, fluffier envelopes are necessarily brighter. Of course, that result is also implied by the explicit expression for L derived from equations (10.6), but the fact that the *product* $L \cdot (\rho_{\text{sh}}/\bar{\rho}_c)$ remains fixed for a given core-mass is hidden by the other dependences.) Furthermore, our theorem, unlike Eddington's (a proportionality requiring normalization), is *explicit*.

However, like Eddington's result (which so upset Jeans and Milne), equation (10.7) contains no reference to *any* details of energy generation. That being so, it must also be more directly derivable, e.g. by inserting the consequences of equation (10.2) into the radiative-transfer equation. This was also quickly verified. Here is the complete, explicit luminosity–compactness–core-mass relationship obtained in these two ways:

$$L \left(\frac{\rho_{\text{sh}}}{\bar{\rho}_c} \right) = \left(\frac{\pi}{6} \right)^2 \frac{ac}{\kappa_e} \left(\frac{G \mu \beta}{\Re} \right)^4 M_c^3. \tag{10.9}$$

Note that in the usual, homology-based derivation of the (unnormalized) classical Eddington luminosity–mass relationship, the term $\rho_{\text{sh}}/\bar{\rho}_c$ disappears 'dimensionally' – in effect, because there is only *one* lengthscale, R_*. The implicit assumption when using homology is that *all* variables of a given type – such as densities – will simply scale together. The important lesson, dramatically underlined by the specific Theorem 2, is that this is *not* true in core-envelope stars.

The analysis also implies that what was previously a (total) mass–luminosity relationship on the main sequence changes over to a core-mass–luminosity relationship for red giants. This becomes more directly obvious when the ratio

$\rho_{sh}/\bar{\rho}_c$ is evaluated explicitly, slightly later in equation (10.10). The result is a steep power relationship between L and M_c for low core-masses, which becomes almost linear for the form of the theory applicable to very massive cores. This explains theoretically another feature of stellar evolution, previously 'proved' by familiarity with computed results: that the luminosities of stars of very different masses, so different in their main-sequence phases, ultimately converge onto what is almost a unique function of core-mass. As far as its luminosity is concerned, a late-stage star with a well-developed core 'forgets' what total mass it started out with.

I turn now to a discussion of Theorem 2 and its generalizations.

Theorem 2 and its modest generalizations

It may initially seem most natural to express the asymptotic relationships found in the above manner in terms of M_c and r_{sh} (or R_c). However, for several reasons already indicated, the relationship between ρ_{sh} and a certain pair of independent variables – $\bar{\rho}_c$ and M_c – is of particular interest. (These reasons include the motivation to examine mass-storage in the first place, the resulting discovery of the importance of $\rho_{sh}/\bar{\rho}_c$ in setting the scales of both the constant τ and the luminosity L, and an interest in the consequences of making a *given* core denser, independently of whether Nature will permit that in practice. Certainly it can hardly be denied that the consequences of evolution making a given central region denser gave rise to this long-standing 'giant' puzzle or paradox.)

Theorem 2 (equation (10.8)) and its analogues for general η result from eliminating the separate dependences of ρ_{sh} and $\bar{\rho}_c$ on r_{sh}.

The best overall value for the temperature exponent η in CNO-burning is 15. Thus our Theorem 2 demonstrates that ρ_{sh} and $\bar{\rho}_c$ not only anti-correlate, but, for all practical purposes, are related in a directly inverse manner, as far as their dependence on each other for a given M_c is concerned. They therefore behave like a child's seesaw: as one goes up, the other goes down. This means that not only are ρ_{sh} and $\bar{\rho}_c$ *not* homologous, but that for a given M_c they may indeed be described as *anti*-homologous.[5] *This* encapsulates the very essence of

---

[5] We now see how singularly inappropriate the term 'shell homology' really is. Its name is not its only defect. First, 'shell homology' starts from an untestable power-law assumption, with no parameter in it like $\rho_{sh}/\bar{\rho}_c$ to determine how close to such behaviour the conditions might be. Second, because nothing like $\rho_{sh}/\bar{\rho}_c$ is identified as having any significance, no-one was ever stimulated to find theorems like that obtained here. Third, it is a completely unnormalized approach – resort has to be made to previously obtained computer models to normalize its consequences for a given composition. This even extends to the task of finding β – the ratio of gas to total pressure used in the equation of state by shell-homology proponents – in a bizarre process that involves integrating a separate equation for the change in β for cases other than the normalizing model. And fourth, it only characterizes what happens to shell variables; there is no direct structural implication such as I find below for the extent of, and mass-storage in, the surrounding radiative region.

giantdom – that envelopes become sparse or fluffy as cores grow denser. The additional dependence upon a fairly high inverse power of M_c, as the latter also grows, only enhances the effect. If ρ_{sh} decreases as both $\bar{\rho}_c$ and M_c increase, then $\rho_{sh}/\bar{\rho}_c$ decreases even faster. It is worth noting this explicitly by re-expressing equation (10.8) in terms of its implications for $\rho_{sh}/\bar{\rho}_c$:

$$\frac{\rho_{sh}}{\bar{\rho}_c} = \frac{g(\#s)}{\epsilon_0^{1/3}\bar{\rho}_c^2 M_c^{8/3}} \,. \tag{10.10}$$

If r_{sh} (or R_c) should remain essentially constant (as is in fact the case; see Section 10.7), $\bar{\rho}_c^2$ in the denominator on the right-hand side of equation (10.10) is proportional to M_c^2, showing more directly how strongly $\rho_{sh}/\bar{\rho}_c$ diminishes with increasing values of M_c. Consequently, using Theorem 1, L itself ends up as a specific and high-powered function of M_c. (For $\eta = 15$, that power is $\frac{23}{3}$; thus L is almost $\propto M_c^8$.)

Thus denser cores induce both fluffier envelopes and larger – sometimes very much larger – luminosities.

What about other values of η? For η near 15, the general results and implications are hardly any different: for each change in η of ± 1, the powers of $\bar{\rho}_c$ and M_c change by only $\pm\frac{1}{9}$ and $\pm\frac{2}{9}$. It is still both remarkable and gratifying that the 'best' value for η yields the density-seesaw property in its simplest and most compelling form.

Before leaving this discussion of the density-seesaw theorem and its equivalent implications for the compactness ratio $\rho_{sh}/\bar{\rho}_c$ (equation (10.10)), I re-emphasize the following point: it is not *just* finding a very low ratio $\rho_{sh}/\bar{\rho}_c$ alone that necessarily gives a red-giant structure – that particular value could exist in a model which stored very little mass beyond the shell – it is the fact that this value both follows from and is intimately coupled with the temperature condition (10.2) involving $\tau \geq 3\rho_{sh}/\bar{\rho}_c$. Good enough though it is to use the approximation $\tau = 0$ to obtain the first (very close) approximation to $\rho_{sh}/\bar{\rho}_c$, that latter value must be fed back to obtain a non-zero value for τ that will lead to the non-negligible mass storage in a red giant's envelope.

10.3.5   Bifurcation and its consequences in the (log ρ, log T) plane

The relationship embodied in Theorem 2 implies a bifurcating behaviour in the conventional (log ρ, log T) plane. Figure 10.2 shows predictions and corresponding results for models of an evolving star of mass $0.8\,M_\odot$, computed using Peter Eggleton's generously provided code. To guide the reader's eye, two asymptotic straight lines are superimposed upon this plot of the computed (log ρ, log T) relationships for models of different core-masses. Since r_{sh} will be shown to be remarkably constant for evolving red giants, these lines reflect

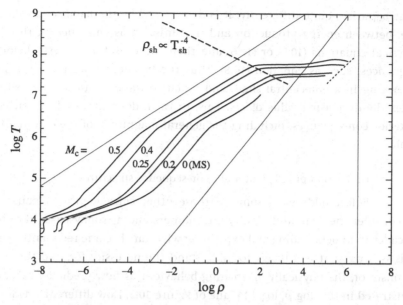

Figure 10.2 Predictions (dotted or broken lines) and corresponding computed results for models of an evolving star of mass $0.8 \, M_\odot$. Core-masses M_c are indicated in units of $M_\odot$.

the consequences of using M_c as a parameter for ρ_c (*not* $\bar{\rho}_c$) and ρ_{sh}, with the further fair approximation that $T_c = T_{sh}$. With these approximations, $\rho_c \propto M_c^2 \propto T_c^2$ and $\rho_{sh} \propto T_{sh}^{-11/3}$. (Figure 10.2 actually shows the negative integer value -4 for the latter power, reflecting an earlier use of $\eta = 16$ for purposes of illustration.)

Simple homology arguments show that during gravitational-contraction phases and early main-sequence stages, the central density and temperature of a star trace out a track of positive slope, $\rho_c \propto T_c^3$, in the $(\log \rho, \log T)$ plane. That is because there is only one lengthscale, R_*. When it exists, the nuclear centre is one (or almost one) with the geometrical centre. However, as our theoretical conclusions and the figure amply demonstrate, the movement of the nuclear centre away from the geometrical centre ultimately creates a bifurcation in this plane. There now result *two* lengthscales: R_c and R_*. They are not completely independent, but involve a subtlety to which we shall shortly turn – the fact that in general the rôle of the radiative zone is *not* to store mass, but rather to enable a convective storage attic to be accessed. Consequently, the overall size of the star, R_*, involves two major components: the relative sizes of these two distinct zones which necessarily lie beyond the nuclear-burning shell.

R_c is of course very directly involved in setting the scale of such quantities as $\bar{\rho}_c$ ($\sim 3 M_c / 4 \pi R_c^3$), and (as r_{sh}) it helps determine T_{sh} too. Thus $\bar{\rho}_c$ is the most natural dimensional quantity given in equation (10.8). It is essential to realize that ρ_{sh} arises quite differently: it is determined by the balance, obviously stable, between energy production and transmission as embodied in the directly physical equations (10.5) or (10.6) and their more useful but less obvious consequences, such as equation (10.8). Thus the balance between nuclear physics operating in a non-central shell, and radiative transfer outside that shell, dictates the decreasing value of ρ_{sh}. This, in turn, determines all the succeeding exterior consequences, including the general magnitude of the external lengthscale.

Unification of red-giant and main-sequence structures

Before addressing some of those other consequences, I pause briefly to consider the general difficulty astronomers and astrophysicists have had for decades in understanding and explaining why stars become red giants. Basically, this is because it is difficult to understand the necessity for, and the moment of onset of, the physically bifurcating behaviour as now predicted here, and as illustrated in the (log ρ, log T) plane of Figure 10.2. How different it would have been had we lived in a time-reversed universe. In the (log ρ, log T) plane plotted in that universe we would see the tracks of ρ_{sh} and ρ_c versus T, representing the progress of two necessarily distinct physical points, approaching one another, not diverging. We would then have appreciated that there were two distinct, if subtly related, lengthscales. With these two points on a collision course in the (log ρ, log T) plane, they would clearly be destined to merge and become one. Only one lengthscale could then survive. Thus, the corresponding problem in that universe: 'Why do stars become red dwarfs?' would be much easier to understand. As easily confirmed, when that merger occurs, the consequence of representing both densities in equation (10.8) by the same simple dimensional expression, involving only one mass M and one lengthscale R, is

$$R \propto \epsilon_0^{1/18} M^{7/9}. \tag{10.11}$$

When shown this merging result, my sober and insightful sabbatical colleague Hans Ritter immediately recognized it (even in a Garching Biergarten) as precisely the main-sequence relationship for the physics employed. Thus, in a time-reversed universe, when the distinct core of a red giant disappears, what results is a main-sequence star. This leads to an inevitable conclusion:

Main-sequence stars are merely degenerate red giants!

Radiative or convective mass storage? – the truth revealed

Radiative mass storage close to the nuclear-burning shell is still very ineffectual. Casual readers may wish to skip the next paragraph, which simply argues for the ultimate necessity of a convective attic to store the bulk of the remaining mass.

In relative terms, the incremental mass fraction which can be stored in the radiative zone grows initially only as $3(\rho_{sh}/\bar{\rho}_c)\ln(r/r_{sh})$, for small values of τ/T_{sh}. However, the additional fact that τ exceeds the stated positive minimum critical value (see equation (10.3)) also permits another mass-storing strategy to be brought into play. Even if mass storage in the radiative zone is negligible (though it is not always so, as I shall show), increasing r (and therefore decreasing T), according to equation (10.2), enables the *corresponding* condition for *convective* mass storage to be accessed. This condition has the same form as equation (10.2) with 2.5 substituted for 4, and a new small constant τ_{conv}, where I now temporarily introduce clarifying subscripts. (It is simply impossible to access this condition for significant convective mass storage, starting from a value for τ_{rad} less than τ_{crit} as given by equation (10.3).) Finally, whether or not much mass has been stored so far in the entire radiative zone outside the nuclear-burning shell, truly significant mass storage is now possible in the convective zone. There, local ρ no longer decreases initially as T^3, and therefore essentially as r^{-3} – the reason for the initially logarithmically small mass storage in the radiative region. Instead, it decreases only as $T^{3/2}$, or essentially as $r^{-3/2}$, initially.

Size of, and mass storage in, the radiative zone

The matching of the radiative and convective T–r relationships at the radiative–convective interface, r_i, implies a relationship between the extent of the radiative zone and the two constants τ. Even without an explicit value for τ_{conv} it is clear that

$$r_i/r_{sh} = q(\bar{\rho}_c/\rho_{sh}), \tag{10.12}$$

where q is a modest fraction, typically of order $\frac{1}{5}$. With this estimate of the size of the radiative region, one can now integrate to find an expression for the mass actually stored in the radiative zone.

I shall sketch these results only to the extent that they are needed below. Call the mass in the radiative zone M_{rad}. A simple integration now shows that M_{rad}/M_c equals $\rho_{sh}/\bar{\rho}_c$ times a mixed, dimensionless, largely *algebraic* (not logarithmic) expression. The essential point is that, because of the nature of equation (10.8) and the approximation $r_{sh} \simeq$ constant, one expects a result close to

$$M_{rad}/M_c \propto M_c^{-14/3}. \tag{10.13}$$

It is again quite gratifying that radiative mass storage in the plotted Eggleton models is well fitted by $M_{rad}/M_c \propto M_c^{-4.4}$. Such dependences for M_{rad}/M_c imply a most remarkable prediction for the moment of maximum convective-envelope penetration; see Section 10.3.6 below.

Size of whole star

Asymptotically, as $\rho_{sh}/\bar{\rho}_c$ and thus $M_{rad}/M_c \to 0$, negligible mass is ultimately stored in the radiative zone. The convective zone takes over the task of storing the mass. Its natural size is determined in the first place by its ability to store the rest of the star's mass with conditions at its base delivered to it by the radiative zone. Initially, the incremental mass fraction M_{conv}/M_c grows within the convective envelope as something like $(\rho/\bar{\rho})(r/r_i)^{3/2}$. However, this simple dependence with radius cannot continue. It relies upon the assumption that little mass has been stored so far, and that assumption itself ceases to be valid.

In the end, the size of the whole star ultimately involves storing the required envelope mass in a way that also satisfies the surface boundary condition. That has to be consistent with the luminosity given by Theorem 1 and the joint surface temperature and density conditions allowed by the opacity and the surface gravity (which involves both the mass of the whole star and its radius). As I have discussed already, Hoyle and Schwarzschild showed that extreme assumptions (radiative or convective envelopes all the way) led to severe under- or over-shooting of the correct surface boundary condition. The final solution involves a judicious choice for the precise interface between the radiative and convective regions. This kind of consideration, with initially deepening convective envelopes, determines the self-consistent size of the whole star both as it evolves along the early subgiant branch and when it has reached the red-giant branch proper. Once that has occurred, the steep temperature dependence of the surface H^- opacity severely limits further reductions in effective temperature. With both luminosity and density deep down jointly and self-consistently determined by the temperature- and energy-balance considerations, the value of the radius is ultimately set by the combination of that luminosity and the highly constrained effective temperature.

The chain of argument is obviously much longer than the assertion: 'a giant is large because it is luminous', or its converse: 'a giant is luminous because it is large', and involves many more elements of self-consistency than either of those simple assertions contain. Nevertheless, on overall balance, if I had to choose between these two extremes, I would come down more in favour of the former, since I have shown that high luminosity and large size arise jointly, the one with the other. (They are born together, as my two asymptotic theorems show. Either of these theorems follows from my analysis without first invoking the other.) On the other hand, the second kind of simple summarizing statement suggests that the luminosity somehow follows from a prior-established great size. I couldn't disagree more strongly with that!

Before leaving this general theme of the bifurcated structure of red-giant stars, I should deal with one possible objection some might make: 'You're dealing with the ultimate structure of red giants, certainly, but you haven't really explained or covered the actual transition to those structures as such. So you haven't told us *why* stars *become* red giants.' As I have shown, the changeover from being a main-sequence star to being a red giant involves a change from a homologously behaving structure with only one mass-scale and one radius-scale, to an anti-homologous structure with two very different mass- and radius-scales. In the first case, the scales are solely outer ones. For red giants, there are both outer and inner scales, with the inner scales determining the overall structure of the star. Thus, not only is there a transition in the key structural properties, there is also a significant change in the logic governing that structure, which manifests itself in many ways. Any attempt to describe fully the transition would have to encompass within itself a change in the governing logic. I suspect that it will be a long time before that daunting end is achieved.

10.3.6   *The moment of deepest convective penetration*

As mentioned many times above, ours is an asymptotic theory of red-giant structure. It should best hold true in the limit as $\rho_{sh}/\bar{\rho}_c \rightarrow 0$, and thus $M_{rad}/M_c \rightarrow 0$. Nevertheless, let us have the audacity to suggest that a simple power dependence such as equation (10.13) holds true for all $M_c \geq M_1$, where M_1 is that value of M_c for which the evolutionarily reducing mass in the radiative region precisely equals the core-mass itself. (In other words, in what follows, the 'small' quantity M_{rad}/M_c will start from the outrageously large value of 1.) Then,

$$(M_{rad}/M_c) = (M_c/M_1)^{-\lambda}, \tag{10.14}$$

where λ is expected to lie between 4 and 5, the range used below. The mass at the base of the convective envelope is M_i or $M_{base} = M_c + M_{rad}$. Thus

$$M_{base} = M_c + M_c(M_c/M_1)^{-\lambda}. \tag{10.15}$$

This function is large as $M_c \rightarrow 0$ or ∞, and therefore has an intermediate minimum. Thus the mass of the convective zone M_{conv} has a definite maximum. This defines the moment of maximum convective-envelope penetration.

The core-mass determining the stationary value is given by differentiating the rhs of equation (10.15) with respect to M_c. It is readily found that it satisfies

$$(M_c/M_1) = (\lambda - 1)^{1/\lambda}. \tag{10.16}$$

The function on the rhs of equation (10.16) is remarkably insensitive to the precise value of λ in the given range. (In fact that function itself has its own maximum of ~ 1.321 for $\lambda \simeq 4.591$.) For $\lambda = 4$ its value is $3^{1/4} = 1.316$, while for $\lambda = 5$ it is $4^{1/5} = 1.320$. Consequently, the core-mass for deepest convective penetration is extremely well determined in terms of the normalizing mass: it is $\sim 1.320\, M_1$. The corresponding value of $M_{\rm rad}$ is determined less robustly. It is $(M_c|_{\rm max.pen.})/(\lambda - 1)$, and lies between 1/3 and 1/4 of the critical core-mass; alternatively, it is $M_1 \cdot (\lambda - 1)^{-(\lambda - 1)/\lambda}$, between 0.33 and $0.44 M_1$.

The value of 1.320 for the critical M_c/M_1 is well fitted by the most recent comprehensive models (Weiss & Schlattl 2000), indeed, spectacularly so for the first sequence that Weiss tested, namely that for $M = 0.80\, M_\odot$, $Y = 0.20$, $Z = 0.0001$: the prediction for critical fractional core-mass 0.3564, tabular entry 0.3585.

10.3.7   Red giant wrap-up: the remarkable near-constancy of $r_{\rm sh}$

I have left one important point dangling all along – the remarkable constancy of R_c or $r_{\rm sh}$ for stars along the giant branch. I shall be brief. If the core were merely a cold white dwarf, its radius would obey the classical relationship $R_c \propto M_c^{-1/3}$ in the low-mass limit. However, it is hot, with an approximately constant temperature $T_{\rm sh}$, say, and that affects its outermost, essentially non-degenerate parts. There, we may consider there to be (i) an ideal-gas, isothermal extension, or (ii) some other treatment that allows for a moderate temperature rise. Refsdal and Weigert (1970) employed the former method, although they had to appeal to model normalizations to obtain an absolute value for $T_{\rm sh}$.

Two factors now make the size of the complete helium core, out to the shell, remarkably self-regulating:

(a)   For a *given* $\rho_{\rm sh}$, if $T_{\rm sh}$ is too large, the additional radial extent and therefore the putative $r_{\rm sh}$ is too large. By our analysis this in turn makes $T_{\rm sh}$ self-consistently smaller. The process converges. For a given $\rho_{\rm sh}$, there is therefore a self-consistent solution for the extra radial extent.

(b)   But $\rho_{\rm sh}$ is not a *fixed* given quantity, but rather has a given, absolutely determined, functional dependence on M_c and $r_{\rm sh}$ through equation (10.8). It decreases quite strongly as M_c and thus $T_{\rm sh}$ increase. This tends to *enlarge* the extra radial extent for which the shell density serves as an outer boundary condition, almost nullifying changes in $r_{\rm sh} \simeq 1.8 \times 10^9$ cm.

These two factors, including the latter one which has not been explicitly emphasized in the past, are together responsible for the very weak dependence of $r_{\rm sh}$ (or R_c) on M_c already noted long ago by Schwarzschild (1958). The fairly weak dependence of the cold-white-dwarf radius on core-mass is cut down by an order of magnitude or more. And *that* is why it is analytically permissible, to a good approximation, to ignore variations in r_{sh} along the red-giant branch.

10.4   'Giant' stars with luminous cores

So far, I have confined attention to those stars, classical low-mass red giants, for which it is almost always a very good approximation to take the luminosity of the compact core to be zero. Such stars have degenerate cores at their very centres, even if they have essentially isothermal, ideal-gas helium mantles surrounding those regions out to the position of the nuclear-burning shell. Because the innermost cores are degenerate, the complete central (helium) structures (core and mantle) are as compact as they will ever be, for a given core-mass. Correspondingly, as the density seesaw theorem strongly implies, the entire stars are as large as they will ever be for the same core-mass.

However, lest it be thought that I am somehow claiming that central degeneracy is essential in order for the star to become a red giant, or to have what in general terms one might describe as 'giant structure', let me repeat here emphatically once more that that is *not* essential. What *is* necessary is that the core be *compact* or *dense* in the context I defined earlier, namely that $\rho_{sh}/\bar\rho_c \ll 1$, as in the initial inequality (10.1) with which I started my discussion. Stars with dense cores in that sense must always have at least some aspects of what we call 'giant-like' structure.

The structures of stars with zero-luminosity, degenerate cores are simply the most extreme example of a continuum of structures for stars possessing dense cores. I shall now proceed to demonstrate both mathematically and physically that less dense and/or more luminous cores promote less extreme though still giant-like behaviour.

10.4.1   A more general theory

When the compact core is itself luminous, one must revert to using the full form of equation (10.4) for the envelope–luminosity balance. I repeat it here for completeness:

$$L_{rad} = L_{sh} + L_c. \tag{10.17}$$

The first two terms can be expressed exactly as in equation (10.5). However, it will be useful to introduce a new variable into the discussion of this more general case, namely the classical factor $(1 - \beta)$, which represents the ratio of radiation pressure to total pressure for a mixture of ideal gas and radiation. For the electron-scattering case treated previously, our entire analysis could have been accomplished in terms of the constant value for each core-mass of that classical factor, in addition to the corresponding value of β itself, the latter already having been used and explicitly exhibited in the earlier equations. I chose not to do so earlier, because I wanted to emphasize separately certain results on densities and their ratios.[6] However, I shall now use that factor in this sketch of the generalized theory. Then the resulting equation has the form:

$$\frac{4\pi c G M_c}{\kappa_e} \cdot (1 - \beta) = (\text{consts}) \cdot \frac{r_{sh}^3 T_{sh}^{\eta+6}}{(1 - \beta)^2} + L_c. \tag{10.18}$$

Matters may be further simplified by expressing all luminosities in terms of the natural (envelope) Eddington luminosity (of the core-mass), also expressing the actual core-luminosity, whatever it might be, in those same terms, i.e. putting $L_c = f L_{\text{Edd}}(M_c)$. When all that is done, equation (10.18) leads to the following simple formula for the $(1 - \beta)$ factor – or x for even greater brevity:

$$x = \frac{A^3}{x^2} + f, \tag{10.19}$$

where, when $f = 0$, A^3 stands for all the *explicitly known* constants in an expression for $(1 - \beta)^3$, the β-equivalent of the equation previously obtained implicitly for ρ_{sh}^3 by substituting shell variables on the lhs of equation (10.5).

An important question now is: when $f \neq 0$, how does $(1 - \beta)$ – or x – depend on f? In other words, how is the total luminosity of a giant-like star affected by the presence of a luminous core as that latter luminosity is itself varied, if that is the *only* thing changed in the model specification? (In general, the value of the core-radius depends upon the core-luminosity. Nevertheless, as theorists we have the opportunity to explore the separate effects of possible variations in both r_{sh} and L, and determine their individual implications.) It is of course

---

[6] Indeed, I am now convinced that earlier reliance by Fred and myself on expressing everything about classical red-giant interiors in terms of β or $(1 - \beta)$, over their whole ranges, was largely responsible for us not obtaining the significant results and theorems derived above involving $\rho_{sh}/\bar{\rho}_c$. At the meeting in Cambridge I showed one page from my thesis, largely taken up in Chandrasekharian mode by one massive, master equation in terms of β in various combinations, occupying most of the page. It was analytically impenetrable! Our interests then ceased at the end of 'first ascent' giant branches with $M_c \simeq 0.5 M_\odot$. Nevertheless, a later extension of that master equation to very large core-masses (0.8, 1.0, 1.2 $M_\odot$, . . .) led to remarkably good matches with Paczynski's empirical core-mass–luminosity relationship in later stages of massive-star evolution (see figure in Faulkner *et al.* 1982). Paczynski's relationship is of course entirely *observational* – the observations being of his own and others' computer output. Note that in our earlier equations β had been written where it properly appears, even though, in the classical low-mass red-giant models, it is essentially 1. In such models, the small quantity $(1 - \beta)$ is itself increasing from extremely small values, low on the RGB, as a very high power of M_c. In a plot of it or L vs. M_c, the horizontal axis is closely hugged until the hidden curve finally explodes out of the axis as $0.5 M_\odot$ is eventually approached. That is when the quasi-linear Paczynski relationship takes over from the previously strong power-law dependence.

possible to answer the question posed by simply computing and plotting the solution to equation (10.19). I have done that. However, some insight is once again available to us by seeing how the introduction of even a small f (with core-radius temporarily fixed) affects the value of x.

Let us introduce that small f; let the initial root to the original equation, $x_0\ (=A) \rightarrow x_0 + \delta$ when some core-luminosity specified by f is introduced into the model. Then from equation (10.19) it is easily shown that, to first order in f, $\delta = \frac{1}{3}f$. Thus, when a luminous core is first introduced, of small magnitude $fL_{\mathrm{Edd}}(M_c)$, the *total* luminosity of the entire model increases initially by only *one-third* of that core amount. In other words, the previous value of L_{sh} produced outside a zero-luminosity core is *quenched* by $\frac{2}{3}L_c$ for small added values of L_c. As L_c increases, it increasingly dominates the total luminosity, the shell luminosity tending to zero. So: luminous cores quench shells – a result often noticed in computer output, but not analytically quantified until now.

10.4.2   *Application to the horizontal branch*

It may initially surprise readers that I now intend to apply these ideas to the horizontal branch, in order to relate the properties of stars in that evolutionary phase to those of red giants proper. But there is a simple two-step process by which one can consider a transition to be made from a model with an idealized zero-luminosity degenerate core at the RGB tip to one with the helium-burning, larger sized core of a star sitting on the horizontal branch:

(a) First, increase r_{sh} only, i.e. while keeping $L_c = 0$, from $\sim 1.8 \times 10^9$ cm (RGB; hot degenerate core size) to $\sim 4 \times 10^9$ cm (HB; size of non-degenerate helium-burning core). (One could re-express all the previous results in terms of core-radius rather than mean core-density, but I shall not take the space to do so here.) The result is that with this single change, by our previous work – e.g. Theorem 1, equation (10.7) – the shell luminosity alone would be reduced from the $\sim 2000\,L_\odot$ characteristic of the RGB tip to $\sim 16\,L_\odot$. Call this latter shell luminosity, calculated as though there were a zero-luminosity core, L_{zlc}. It represents a natural 'base' value or lower limit to the luminosity of a horizontal-branch star, were such a star to have the core-radius it normally does have, but with all the core-luminosity suppressed.

(b) Second, add in the consequences of having a non-zero core-luminosity, L_c. The effects, including the quenching of the shell from its base value L_{zlc}, can now be obtained from our analysis in the previous subsection. As an example, if the luminosity of the core itself *is* the previously established value L_{zlc}, then the full luminosity of the star, $L_* \simeq 1.46 L_{\mathrm{zlc}}$.

Now the luminosity of a bare helium core (i.e. a low-mass helium main-sequence star) is comparable to the values just derived for the base HB shell luminosity, L_{zlc}. (This is not the fortuitous coincidence it may at first appear to be.) This means, according to our analysis, that the luminosities of these two parts of an HB star are comparable. Such stars are thus substantially less luminous than their immediate evolutionary precursors, but are also finally in the range where the 'quenching' described previously must take place. The analysis therefore implies that the total luminosities that result are a small fraction of their previous, RGB-tip values, and that they also increase rather modestly with Z_{CNO} on the HB itself. Meanwhile, although the radii of the resulting stars are correspondingly less than those of their immediate precursors, they too increase, but much more sharply with increasing Z_{CNO} on the HB, resulting in substantially cooler stars for 'high-metal' HB stars than their 'low-metal' counterparts. (The latter effect is of course what I myself discovered computationally – Faulkner (1966). However, the referee (Martin Schwarzschild) was justifiably but kindly somewhat dubious about the handwaving explanation I offered in that paper.)

Those computing evolving stellar models have become very familiar with the result that ignition of a new central fuel quenches a previously burning shell while *reducing* the overall luminosity of a star. They have seen it many times, in fact so many times that it simply bores them. Indeed, like the 'node', 'concertina' or 'mirror' 'principle' (sic) for the constancy of a shell source's radial position (another result where repetitive reproduction substitutes for understanding), no explanation is attempted for this result since 'it always happens, anyway'. However, as far as I am aware, the arguments now provided above represent the first time that an answer has been given to a certain question which may never even have been specifically asked before, but is in a sense generic for stars with a new central fuel ignited in previously non-burning dense cores: 'Why are HB stars so much fainter than their immediate RGB-tip precursors with the same mass, core-mass, and overall composition?' Now the reader might well say, 'Hang on a minute, how do you even know that the theory applies to HB stars?' The answer is that it applies to *any* type of star with a sufficiently dense core, i.e. one for which the resulting value of $\rho_{sh}/\bar{\rho}_c$ is small enough when the theory itself (including the energy generation law) is applied.

Unfortunately, Nature has played a slightly disappointing trick on us, at first sight. While the values of r_c or r_{sh} are small enough for the theory to apply well to most red giants, and particularly so for those at the RGB tip, this is not so for all HB stars. In fact, by the very token of reducing the shell luminosity to *a maximum* of L_{zlc}, with the reduction factors suggested above, or equivalently by our new Theorem 1, the reduction of L_{sh} from RGB values for a given M_c necessarily occurs because $\rho_{sh}/\bar{\rho}_c$ *has* increased – by the very same factor! Consequently, we find that the values of $\rho_{sh}/\bar{\rho}_c$ for high Z_{CNO} HB stars are $\sim$0.04–0.05, i.e. in a range where we know the theory is at best only roughly true, and at worst, breaking down. And, to the extent that anything about it can still be trusted, our density-seesaw Theorem 2 shows that the situation only gets worse for models with low Z_{CNO}.

On second thoughts, however, this is not such a bad situation; it corresponds precisely to where such stars are both seen and computed to be, in relative terms. Suppose for a moment that the naturally occurring r_c or r_{sh} were lower by merely a factor of $\sim$3 for both RGB cores and for HB cores. Both evolutionary phases would then satisfy our theory well, the RGB exquisitely so, and the HB models as well as or better than the RGB does now. *In that case we would see virtually no horizontal branch.* Virtually all HB stars would be jammed onto the RGB we now see (if necessarily much more luminous and thus close to the current RGB tip), while the RGB stars would evolve to much higher luminosities. Restricted by the surface boundary condition, the RGB stars would still occupy a region looking like a continuation of the classical RGB, such stars now being permitted to evolve on to much higher RGB luminosities, provided that increased envelope mass loss did not substantially limit that further climb.

So we can now belatedly recognize that the very existence of the horizontal branch as we know it is owed to the happenstance that the asymptotic theory of the RGB breaks down when applied to HB stars precisely for the range of Z_{CNO} provided by Nature – that is why HB models in which *only* Z_{CNO} is varied, in the observed range, occupy positions in the HR diagram which cross the lower region of the classical Hertzsprung gap. Because of the self-consistent conditions provided by putting high Z_{CNO} into the theory, we end up with current, high-metal HB stars in our specific universe hardly distinguishable from true red giants in their exteriors (they both need and have large mass-storing convective envelopes), while low-Z_{CNO} models have such large values of $\rho_{sh}/\bar{\rho}_c$ (even a factor of $\sim$2 can make all the difference in this range) that they don't need convective envelopes, can make do with entirely radiative ones, and for that reason end up with hotter and bluer surfaces.

10.5   The horizontal branch as missing link

For a given value of Z_{CNO}, the zero-age horizontal branch may be described by a sequence of models in which a variable amount of envelope mass is stored above a helium-burning core of given mass. A vanishingly small envelope mass leads to extremely blue, radially small models; while a large envelope mass, comparable to the core-mass, leads to red, i.e. large models. Furthermore, the higher the Z_{CNO}, the redder the models – until they run into the the RGB where the outer boundary condition and the more efficient convective envelope mass-storage prevent further redwards extensions of such sequences. All this was broadly noticed long ago (Faulkner 1966), but I confess that the importance of the horizontal branch as a transitional structural type between 'normal' and 'giant' *hydrogen-burning* stars does not seem to have been recognized before, either by me or anyone else. (To be quite clear on this: I did see HB stars with low envelope masses as 'transitional', but to helium main-sequence stars, which result when envelope masses become negligible, and not to other hydrogen-burning stars.) This is really rather ironic. The problem of understanding the evolution of intermediate or more massive stars from the main sequence to the giant branch can be posed as the question: 'Why do stars developing denser cores cross the Hertzsprung gap?'

One thing in particular gets in the way of understanding and answering this question: namely that for massive stars there exists no type of long-term static structure occupying this region for the large core-masses generated by the end of the previous, centrally convective main-sequence stage. Yet for lower masses, Nature has provided us with a series of static structures, with helium-burning cores, that do occupy and serenely march across that very domain – the horizontal branch. In this section I shall show how one can exploit Nature's bounty in that regard to provide insight into this very old question. I shall show that certain horizontal-branch models, when induced to change internally in an appropriate way, behave *predictably* and in every way in a manner entirely analogous to the behaviour of more massive stars as they evolve across the Hertzsprung gap. In this way, the horizontal branch acts as an evolutionary *missing link* between the slow, steady and continuous evolution of low-mass stars to the red-giant branch, and the more rapid, out-of-equilibrium evolution of more massive post-main-sequence stars. For the most part, such stars are out of thermal equilibrium, but not thermally unstable as Renzini has claimed, alone or in concert, in a series of papers, e.g. Renzini *et al.* (1992).

10.5.1   *Marching across the Hertzsprung gap*

As implied above, the classical ZAHB (zero-age horizontal branch) for a given Z_{CNO} consists of stars essentially in complete quasi-static equilibrium, in which very largely just one parameter, the envelope mass, varies. As such, those ZAHB models do not march across the Hertzsprung gap, rather they stay forever in place. However, we may construct other extremely useful if imagined zero-age 'sequences', which not only themselves also span the lower reaches of the Hertzsprung gap but which then lead on to a very interesting and significant numerical experiment, which strongly illuminates the whole question of the behaviour of intermediate- and high-mass stars in their post-main-sequence evolution. We shall find that the very features claimed by Renzini to support his own explanation for the transition to red giants, particularly of intermediate-mass stars, are instead a *consequence* of changes occurring in the deep interior. (Renzini suggested a cooling/recombination/increased-opacity cause for thermal instability in the star's outermost layers, an attempted explanation which I shall critically dissect further, below.)

I discuss new sequences of low-mass HB-like models that do march across the Hertzsprung gap from left to right. The point is this: consider a different possible sequence of ZAHB models, in which the one varying parameter is simply Z_{CNO}. (Yes, both the envelope and the core-masses are held constant in this exercise.) As Z_{CNO} is successively increased in such ZAHB models, the positions they occupy in this 'sequence' move from left to right across the Hertzsprung gap, with a moderate increase in luminosity (Faulkner 1966). Because of the presence of the essentially unaltering helium-core luminosity, and the shell-quenching effect discussed earlier, the previous red-giant luminosity dependence for other given parameters – i.e. $L \propto \epsilon_0^{1/3}$ – is cut down to roughly $L \propto \epsilon_0^{1/6}$, as shown very well by the solid line in Figure 10.3.

This figure was constructed some years ago when Eric Sandquist (then a UCSC graduate student) and I were exploring various aspects of HB and related evolution, and I realized that the HB could act as the missing link now under discussion. In order to construct this diagram, we used Peter Eggleton's extremely versatile stellar evolutionary code. Because of its very general kind of construction and solution method, including a 'floating mesh', it is often possible to perform ingenious numerical experiments that would either defeat, or at least provide difficulties for, a more conventionally written code.[7]

---

[7] I am sure that Peter Eggleton would be the first to acknowledge that this method goes back to an extremely insightful hydrodynamics seminar by Douglas Gough, the present successor to Fred Hoyle as Director of the Institute of Astronomy, when he was either still a graduate student, or possibly a very new post-doc. I still recall the frisson of excitement and insight I then experienced as Douglas packed his points into the two boundary layers, where they were needed and would do the most good, using an automatic algorithm which simultaneously solved the problem of minimizing an estimation of the presumed next-order errors in the integration scheme. 'Brilliant!' I then thought, and do to this day. In Peter's hands, Gough's method, allied to the use of the variables I first introduced in about 1961 and employed in my Cambridge Ph.D. thesis under Fred (Faulkner 1964) were used to great effect. For the record, those latter variables – called 'Faulkner variables' by all of Fred's students – exploited the fact that appropriate powers of the integration variables of interest would all have natural expansions in terms of r^2 near a stellar centre. Consequently, the old, inaccurate and now superfluous 'central stellar core expansion' could be dispensed with, and a more accurate boundary integration readily performed.

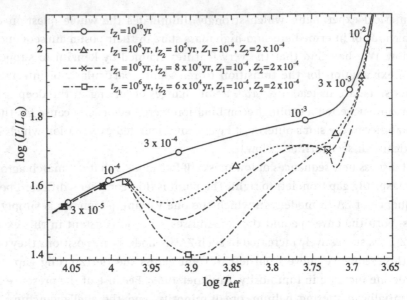

Figure 10.3 The effects of increasing Z_{CNO} in horizontal-branch models. Solid line/circles: 'static' models labelled by their Z_{CNO} shell-source content; remaining lines/symbols: tracks followed by evolving models in which time-varying increments in Z_{CNO} are imposed. (Specifications described in text.) All sequences start with $Z_{CNO} \simeq 10^{-5}$. The more 'dipping' tracks correspond to faster timescales for Z_{CNO} increments after reaching 10^{-4}. Models indicated by their respective sequence symbols (triangle, cross, square) in the lower middle of the figure all have $Z_{CNO} = 10^{-3}$. They are 'lagging' in their expansion to the corresponding ZAHB $Z_{CNO} = 10^{-3}$ position and radius, their envelopes absorbing energy as they attempt to reach that ultimate static state. Conventional post-ZAMS evolutionary models with sufficiently rapidly evolving cores exhibit analogous behaviour for similar reasons.

The upper solid-line sequence of ZAHBIC models (ZAHBIC ≡ Zero-Age Horizontal Branch with Increasing CNO), shown slowly increasing in L across the diagram, was obtained as follows. The value of Z_{CNO} in ZAHB models made with Eggleton's code was very slowly increased from its initial value of $\sim 10^{-5}$, on a 10^{10}-year timescale. This timescale was so slow that these were certainly static models, in effect. (Of course, real stars spend only some 10^8 years at most in the entire HB domain before they scurry back to the asymptotic giant branch (AGB) with exhausted helium cores. So, in order to generate this sequence of models, the composition variables were frozen at their original values, thereby producing a true ZAHBIC sequence – another very useful possibility, controlled by just one input parameter in Eggleton's code.) Note that whereas models with $Z_{CNO} \simeq 10^{-4}$

have $T_{eff} \simeq 10^4$ K, and are therefore 'blue', those with $Z_{CNO} \geq 3 \times 10^{-3}$ are much cooler, definitely 'red', and virtually jammed onto the RGB. They correspond precisely to certain models which the discerning eye can pick out from the figures in Faulkner (1966).[8]

The remaining (dotted or dashed) curves show what happens when the following controlled changes in Z_{CNO} are imposed. Starting from $Z_{CNO} \simeq 10^{-5}$, the value of Z_{CNO} is increased with a certain imposed pattern of e-folding timescales. Up to the value $Z_{CNO} = 10^{-4}$, the e-folding timescale is 10^6 years. This timescale, though fairly short, is still a fair bit longer than the natural thermal timescale for a disturbed envelope to return to equilibrium. Thus, the HR diagram trajectories for all these additional models with $Z_{CNO} \leq 10^{-4}$ are not only the same in this region (timescales for changes being identical so far), but also they are little removed from the corresponding ZAHBIC models.

Significant differences appear, however, when the timescales for increments in Z_{CNO} are successively and smoothly decreased to the three values of t_{Z_2} indicated in the figure. The reductions in timescales were controlled as follows. Starting at $Z_{CNO} = 10^{-4}$, the e-folding timescales were reduced sinusoidally to become set at their new values once Z_{CNO} reached 2×10^{-4}. The timescales were then kept at those new values thereafter. (This explains the meaning of the non-obvious lettering littering the diagram.) The new e-folding timescales reached after $Z_{CNO} = 2 \times 10^{-4}$ years (and $10^5, 8 \times 10^4, 6 \times 10^4$ years, respectively) *are* shorter than the envelopes' thermal timescales. The dramatic 'luminosity-dipping' behaviour now exhibited by the Z_{CNO}-changing trajectories is testimony to the fact that the envelopes have been driven out of equilibrium by their response to the carefully controlled and deeply imposed changes.

It is particularly worthwhile to study the implications of all the models having $Z_{CNO} = 10^{-3}$ in these timescale-changing sequences. In thermal equilibrium, of course, all such models would be identical, and positioned at the 10^{-3} point circled on the solid-line ZAHBIC sequence. However, as the incremental timescale reduces to smaller values, the corresponding out-of-equilibrium models in such successive trajectories are found to be both lower in L and higher in T_{eff}, so that the time-changing radii are successively so much smaller than they ultimately need to be in a fully relaxed, equilibrium ZAHBIC model sequence.

This is the key to understanding not only the behaviour of these models, but also the long-computed, characteristic surface behaviour of all evolving post-main-sequence intermediate- and high-mass models.

---

[8] Unfortunately, not all eyes or verbal descriptions have been so discerning. Not only are the theoretical ZAHB figures 'after FAULKNER, 1966' in Kippenhahn & Weigert (1990) extremely poor, but the description of what is illustrated confuses core and total masses. This produces a quite unbelievable end result.

What, then, is going on? As can be seen by studying the initial ZAHBIC models, the outer radii of the equilibrium ZAHBIC models become larger and larger with increasing Z_{CNO}. This is particularly so for Z_{CNO} in the range of 10^{-4} to 10^{-3}. This is because such envelopes are entirely radiative. Therefore, they behave just like either the polytropic or the completely radiative equilibrium envelopes treated in Appendix 10.A. Although the sensitivity of their radii is somewhat reduced by virtue of the presence of a helium-burning core and the consequent quenching effects discussed previously, such models are simply analogues of the extremely sensitive low-mass post-main-sequence transitional results obtained by Sandage and Schwarzschild (1952). (Recall that the latter computations were what resulted when the possibility of convective envelopes was not included in the computations.) As Z_{CNO} is increased further, and the RGB is closely approached, however (and for the very same reasons), the job of envelope mass-storage is taken up by a convective outer envelope, which reduces drastically the sensitivity of the outer radius to changes occurring deep below.

(I have always found it helpful to think of the demonstrated sensitivity of both low-mass RG and HB radii as follows. Imagine everything to be permitted to be just radiative envelopes, to begin with. At some critical stage, as indicated by my general theory, the radii of mass-storing envelopes become extremely sensitive. For small changes in the governing deep parameter, radii increase by huge factors while luminosities increase much more modestly. Were such envelopes permitted in Nature, the radii would be simply huge, and effective temperatures could be as low as literally a few tens of degrees. However, the extremely efficient mass-storage capability of outer convective attics changes all that. We may imagine a line or curve of grossly changing overall surface opacity in the HR diagram, essentially parallel to the observed RGB, moving in from such very low values of T_{eff}, and sweeping up every representative point in its path. The most extreme of the earlier, naively calculated radiative envelopes then become convective envelopes on the classical RGB. Models already hotter than some critical amount are essentially unaffected by this imagined process, while some, whose initially imagined, fully radiative envelopes would have placed them in the general vicinity of the ultimate RGB already, are moved to only modestly hotter positions. This qualitative analysis of the general jamming up that is to be expected accounts for all the kinds of effects seen in old and more recent RG computations and in HB results.)

In the original ZAHBIC models, as Z_{CNO} is increased, not only do the outer radii increase substantially but the radial position of main mass storage becomes more and more removed from the dense core, for all the reasons given before. This means that such successive equilibrium envelopes necessarily possess much more energy, gravitational plus thermal, *in toto*. However, although the increase in Z_{CNO} in the time-changing models temporarily increases the nuclear-burning rate substantially, the latter relaxes back fairly quickly to more modest levels as the densities in the shell and in region 3 decline. This has several consequences for the models' overall energy budgets, their resulting behaviour, and our ultimate understanding of what is really going on both in these specially constructed examples and in the post-main-sequence expanding evolutionary developments in essentially all stars:

(a) Initially insufficient additional nuclear energy is generated to provide the new total envelope energy needed by the equilibrium ZAHBIC model for the corresponding new value of Z_{CNO}. Furthermore, although additional energy *is* being generated in a Z_{CNO}-changing model, it is deep down inside the star, in the shell at the base of region 3, and therefore not where it can do any good for the mass-storing envelope's immediate energy budget requirements. Thus, the envelope does not have immediate access to the energy needed to bring it to an ultimate destination corresponding to the new current value of Z_{CNO}.

(b) Nevertheless, as with any ultimately stable physical situation, the entire system still attempts to seek the solution corresponding to the ZAHBIC equilibrium model. In order to do this, while having insufficient energy initially, the envelope either expands only very slowly (for the longer e-folding timescales), or not at all (see the initial slope of the dip for the shortest timescale indicated in the figure), or, one suspects, might even have shrunk modestly were the imposed timescales to have been even shorter. (I remark that later, again for the shortest timescale shown but at larger values of Z_{CNO}, the envelope finally reaches its own RGB-like position with an overall radius *smaller* than immediately before. The curves here show the influence of growing convective envelopes.) Nevertheless, whichever of these detailed surface developments is occurring, initially the envelopes remain much smaller than required in the corresponding ultimate equilibrium ZAHBIC models, and the more so the faster that Z_{CNO} is being forced to increase.

(c) In order to obey the expansion imperative that is still operating in region 3 at the base of the envelope, the envelope ends up taking energy from the one resource remaining for it: the luminous energy coursing through the envelope itself. *This* is what is responsible for the deeply dipping surface luminosity, which becomes so much more pronounced as the Z_{CNO}-increasing timescales themselves decrease.

(d) Thus, in the end, we have the following two consequences for the time-changing models: (i) they are much smaller 'than they would like to be', i.e. than the corresponding ZAHBIC models; (ii) they are a fair amount less luminous 'than they would like to be'.

(e) Thus the envelopes are in fact *lagging behind* in the properties they would ultimately acquire had they only been able to come to the relevant equilibrium. The needed expansion towards the latter goal is what sends the envelopes out of equilibrium, creating an appetite for energy that can be met only by the local luminous flux.[9]

(f) All the arguments I have been making here for these specially constructed ZAHBIC models and their time-developing cousins apply with equal force, *mutatis mutandis*, to evolving and expanding post-main-sequence models of essentially all stars. We now see that it is most definitely *not* an envelope thermal instability that drives the expansion of the entire star, as Renzini, either alone or with a succession of (sometimes absconding) colleagues has long asserted. To the absolute contrary, the envelopes are sent out of equilibrium by an expansion imperative dictated by developments in the deepest and densest interiors of the stars.

(g) And so, in a familiar phrase, we arrive at something we have seen and noticed all along, but truly know it now for the first time: that stars expand, and do so mightily, when dense cores develop in their deep interiors.

10.6   All together now: a unification of immediate post-main-sequence evolution for all masses

The theory and ideas described in Section 10.5 may be applied to stars of greater stellar mass, as they themselves develop dense cores and leave the main

---

[9] The points made in these last few lettered items correspond almost exactly with my earlier explanation for the behaviour of largely radiative, mass-losing stars (Faulkner 1976). (Generically, convective stars behave differently.) Ron Webbink has generously acknowledged that that work, initially begun at the Institute of Astronomy during a 1973 sabbatical leave, set the scene for his own outstanding thesis and later research by him and many others. In that early to mid 1970s work, various *imposed* mass-losing timescales resulted in radius- and surface luminosity-reducing behaviours which also grew stronger as timescales were reduced. By comparing such radius behaviours in particular with the Roche-lobe sizes of correspondingly mass-changing lobe-filling binary members, the actual timescales for mass transfer out of a lobe-filling star to *any* companion became a simple eigenvalue situation. However, the fundamental *reason* for the observed computational behaviour could be traced to the overall, masspoint-by-masspoint energy differences between initial equilibrium models and 'target' equilibrium models of slightly reduced mass. This was once again the first detailed explanation for what had also become frequently observed behaviours of computed models, seen so many times that people believed that it had been understood: that stably mass-transferring radiative stars do so on the overall thermal timescales of stars.

sequence. The specific character of their cores may be different (see below), but the key point is that the very principles discussed earlier apply here too: that is, that, following hydrogen core exhaustion, a period of gravitational contraction ensues in which a denser core naturally develops, and hydrogen burning shifts, perhaps through a thick shell-burning phase, towards a thin shell external to that core. For reasons described all too cursorily in the next paragraph, such cores experience faster changes, and are initially heated through gravitational contraction. There naturally result stars with small but not initially minute values of $\rho_{sh}/\bar{\rho}_c$ and non-negligible luminosities in non-degenerate helium cores – twin factors making them analogues to that extent of horizontal-branch stars. Because more massive stars in the immediate post-main-sequence phase share these twin factors with horizontal-branch stars, the same kind of structural treatment naturally applies to their envelopes (with one significant caveat). Consequently, such stars necessarily take up positions in the same general area of the HR diagram, albeit at higher luminosities because of their larger core-masses: the region generally known as the Hertzsprung gap. We return to the caveat in comparing them with HB stars a little later.

In slightly greater detail, there are of course important qualitative differences between more massive stars and low-mass stars as they leave the main sequence.[10] Such differences largely reflect the mode of hydrogen-burning nuclear energy production. The main-sequence cores of intermediate- and high-mass stars are convective because of CNO-cycle energy production, and, through the convective mixing, they are homogeneous up to the point of hydrogen exhaustion. That exhaustion occurs essentially simultaneously over the entire core, producing a very large, non-burning helium core, compositionally differentiated from the rest of the star. The core-mass fractions of such stars exceed the Schönberg–Chandrasekhar limit (Schönberg and Chandrasekhar 1942), so that the longer timescale allowed for cores for lower mass to become essentially isothermal and then gradually enter the nearly isothermal, electron-degenerate domain of pressure support is denied them. Faster contraction ensues, releasing significant amounts of gravitational energy that both heats the core and results in a substantial core-luminosity at the same time, while the thinning hydrogen shell undergoes readjustment. More rapid core contraction, ultimately terminated by helium-burning ignition, gives a much faster timescale for internal developments than in the case of lower-mass post-main-sequence stars. That fact is responsible for the relatively greater speed with which more massive stars

---

[10] The broad-brush account in this paragraph necessarily leaves out many fine details of the evolutionary developments that stars make when leaving the main sequence. Such details are treated exhaustively for a model of five solar masses in a classic review by Iben (1967); references to similar studies for other masses may be found in that typically masterful review.

evolve across the HR diagram in response to these changing internal conditions, and is the fundamental reason for the very existence of the Hertzsprung gap itself.

Nevertheless, to repeat the point made earlier, these more massive stars share with their lower-mass horizontal-branch counterparts the properties of having increasingly dense helium cores (if not so dense as low-mass degenerate cores) and non-negligible core-luminosities. The major factor now differentiating the more massive stars from their conventional HB counterparts is the relatively short timescale for their internal developments, as their cores contract gravitationally. In that respect, the more appropriate comparison is with the HB models previously driven out of equilibrium by the rapidly imposed timescale for changes in CNO content.

Although the specific causes for internal change are different, the consequences are observationally similar. Changing conditions deep down, involving dense cores, require envelopes for potential new equilibrium models to be successively larger. In the case of the HB models driven by increasing CNO content, the 'target equilibrium model' for a given amount of Z_{CNO} is fixed, the radius larger for greater CNO content. In a certain range the sensitivity of radius to CNO content is quite strong, because these models have radiative envelopes. Such a behaviour mimics the strong sensitivity found by Sandage and Schwarzschild (1952) before the advent of convective envelopes in evolutionary computations. For more massive stars evolving off the main sequence, the 'target equilibrium model' also becomes larger as the core grows denser – the same overall general relationship between dense cores and fluffy envelopes explored throughout this paper.

But now comes the important caveat mentioned towards the end of the opening paragraph of this section, and the reason why the complete phrase 'target equilibrium model' was highlighted in the previous paragraph. That is because for such relatively massive cores, there is no fixed 'target equilibrium model'. The 'target' is always shifting to the right in the HR diagram, the 'target radius' becoming ever larger because of the great sensitivity of external radius to central compactness as the core continues to contract. This process and prolonged chase by the envelope can cease only when the core has reached its own longer-term new equilibrium. In other words, the target keeps on moving.

We have now come to the point where the behaviour of our HB models with changing CNO content as the analogues, or missing link, to the evolution of more massive stars can be exploited. For those evolving HB models, 'dipping behaviour' occurred in the HR diagram when the timescale for central developments became shorter than the envelopes' thermal readjustment time. The envelopes were forced to play catch-up in their attempt to expand to the new target equilibrium. Such dipping behaviour is ubiquitous amongst all post-main-sequence models of intermediate mass, and for the same reason: their envelopes are attempting to play catch-up, and the energy they require to expand to the ever-moving target comes at the expense of the radiation field coursing through it. Thus, such envelopes are not expanding because of some surface opacity effect due to recombination sweeping inwards, as Renzini has maintained; in complete contrast, they are making changes dictated by developments in their deep interiors. The changes are not driven by a thermal instability: the envelopes are changing on their thermal timescales because they find themselves to be too small for what would be a new, time-changing equilibrium, if only they could reach it. They are always relaxing towards that ever-changing, larger 'equilibrium target'. A thermal instability would necessarily produce structures that change on a thermal timescale; but structures changing on thermal timescales may merely be out of thermal balance and not necessarily thermally unstable.

As models become more massive, the post-main-sequence dipping behaviour declines, and ultimately ceases. The reason for this is also fairly clear. At higher luminosities and lower envelope densities, the thermal timescales for such envelopes become smaller at a faster rate than do the central core changing times. As these two timescales come together, the expanding envelopes can now essentially match what is happening internally, and the need to play catch-up and absorb significant energy from the internal radiation field disappears. Envelopes have now been returned to an essentially static role in which they can take up the sizes required by the combinations of shell and core nuclear burning and their deep core sizes at given times.

We can now contrast this with Renzini's proposed explanation for the expansion of stars to the red-giant branch, an explanation he advanced particularly for intermediate-mass stars. Renzini suggested that as a star of intermediate mass starts to expand from the main sequence – a fact taken as a given – recombination occurring in its cooler regions increasing the opacity there, leading to absorption of luminous energy from below. That in turn expands the star's outermost layers still further, leading to more cooling and yet more absorption, as the hypothesized recombination wave sweeps inward through the star. A thermal instability develops from the outside inwards, for which the luminosity dips are advanced as evidence for this absorption of flux in the outermost regions. He has furthermore claimed that this opacity effect is naturally smaller when there are fewer heavy elements in a star, and that this is why such stars do not expand as much. In this explanation, once the initial expansion from the main sequence has made the outermost layers cool enough, the natural behaviour of those layers takes over and promotes the major, fast transition to the red-giant branch itself. Thus the behaviour of the outermost layers *leads* the star to the giant branch.

I have shown above, instead, that the behaviour of the central regions is the main driver. Far from leading the rest of the star along, the envelope regions are *lagging behind* where they would have been had there been time for them to reach either a complete new equilibrium mandated by the change in central conditions (in particular the increasing central condensation) or some 'moving-target' analogue. It is no coincidence that the outer parts of shell-source stars expand when their cores become denser; as I have shown, that is what their new equilibrium (or analogue moving-target) structures demand. As for the lessened tendency of metal-weak shell-source stars to expand (either so vigorously in transition; or so much, finally), that too is easily understood. For given core conditions, such metal-weak stars with radiative envelopes naturally have substantially smaller equilibrium structures than their metal-strong counterparts. At any given stage in their central development, therefore, expanding metal-weak envelopes have less catching up to do, are not so much out of equilibrium, expand more sedately, and come to naturally bluer final equilibrium positions.

It may be worth a small digression here to emphasize this point of extreme sensitivity of size to metal content for such stars. Although long in the literature in a specific context (Faulkner, 1966), it does not seem to be widely appreciated as a fairly general result. In a certain range, the radii of shell-source stars with radiative envelopes are extremely sensitive to their metal content, through the value of Z_{CNO} in their shells. My old horizontal branch models (Faulkner 1966) and the ZAHBIC models shown above illustrate that particularly strongly. Let us examine that dependence in the range of Z_{CNO} from 10^{-4} to 10^{-3}. The ZAHBIC models in Figure 10.3 show that $R \propto Z_{CNO}^{0.565}$ in this range. This is an astonishingly strong dependence for anyone brought up on standard homology expectations![11] Let us see what those standard homology expectations are. The metal content is so low and the (ρ, T) conditions are such that the envelope opacity in HB stars is certainly dominated by electron scattering. We all know that main-sequence stars with electron scattering have luminosities that do not depend at all upon the normalizing energy-generating factor ϵ_0. The radii, however, do depend upon

---

[11] In addition to the inspiration and challenge provided by Hoyle and Schwarzschild, frequent arguments in early 1964 with standard homologist Donald Lynden-Bell and fellow cyclist Russell Cannon were also partly responsible for my embarking on my original horizontal-branch work. In those almost weekly encounters in my DAMTP office, I asserted an expectation (naively argued but happily confirmed) that normal homology would simply not apply to the CNO-shell-source stars that I expected HB stars to be. (The valuable contribution of those arguments in shaping my thoughts was implicitly acknowledged in my 1966 paper.)

that factor, but very weakly. For CNO-cycle, main-sequence stars with electron-scattering opacity, if we take $\eta = 15$, we find $R \propto \epsilon_0^{(1/18)} = \epsilon_0^{0.056}$, as implied earlier in my 'degenerate-red-giant' main-sequence relationship (10.11). Thus, horizontal-branch stars, as an archetype of non-zero-L_c, radiative-envelope shell-source stars, have a surface radius dependence in this range of metal content *ten times greater* than that of main-sequence stars of comparable or greater luminosities. Were L_c zero, this result could be traced immediately to our density-seesaw theorem, equation (10.8), and to the extreme sensitivity to $\rho_{sh}/\bar{\rho}_c$ of radiative envelopes built upon cores as discussed previously and also explored, in effect, in Appendix 10.A.2. With L_c non-zero, however, modest modifications occur to both total luminosity L (via equations (10.17)–(10.19)), and correspondingly to the values of ρ_{sh} or $\rho_{sh}/\bar{\rho}_c$. (*Total luminosity L and ρ_{sh} or $\rho_{sh}/\bar{\rho}_c$ still have an inverse* dependence on each other, as is evident by reference to equations (10.5) and (10.18), or alternatively to the completely separate derivation leading to equation (10.9), which still holds for stars with sufficiently dense cores.) For our HB models, as noted already in Section 10.1, the total L dependence is changed from $\propto \epsilon_0^{1/3}$, appropriate to the zero-L_c case, to approximately $\propto \epsilon_0^{1/6}$. Thus L still increases with ϵ_0 (or Z_{CNO}), though more weakly. The concomitant decrease in $\rho_{sh}/\bar{\rho}_c$ as the factor ϵ_0 is increased in turn leads to a substantial and much larger increase in surface radius because of the much greater sensitivity to shell conditions possessed by dense core/radiative envelope models. This is the basic reason why at these luminosities HB models in which Z_{CNO} is the only varied parameter span almost the entire width of the HR diagram from main sequence to red-giant branch. Yet in a sense, as I hope I have shown, this is all part and parcel with the great post-main-sequence radiative radius sensitivities found by those pioneering investigators Sandage and Schwarzschild, and Biermann and Temesváry (unpublished), in addition to Hoyle and Schwarzschild.

To return to the main point, and to summarize: the envelopes of these expanding stars are lagging behind, not leading the charge. Stars do not become red giants because of a superficial cause sweeping inwards; they do so for much deeper reasons.

10.7   Summary and conclusions

By pursuing to its conclusion an analytical treatment of low-mass red-giant stars based on Fred Hoyle's ideas in the 1950s and 60s, I have been able to explain many properties found previously only in computed evolutionary sequences. I have also made a number of verified predictions. Novel aspects in the asymptotic theory include a 'two-centre' approach in which the quantity $\rho_{sh}/\bar{\rho}_c$ (the compactness, or ratio of shell to mean core-densities) features prominently and is solved for self-consistently. Novel consequences include (i) explicit algebraic theorems such as (a) a relationship between core-mass, luminosity and central compactness, and (b) a density-seesaw theorem anti-correlating shell and mean core-densities, (ii) the lengthscale of the radiative region surrounding the core and its connection to the inner core size, and (iii) a remarkably accurate predictor for the moment of maximum convective envelope penetration. I have also introduced a viewpoint which unifies red-giant, main-sequence and post-main-sequence stellar structure for all masses.

I have shown that, in a sense, red giants are easier to understand explicitly than main-sequence stars, and that main-sequence stars may be profitably viewed as mathematically degenerate red giants in which the core-mass has shrunk to zero. I obtain classical main-sequence proportionalities when that is done. On the other hand, I have applied these same ideas to models for horizontal-branch stars, and have shown that they act as a missing link to the post-main-sequence behaviour of even more massive stellar models.

In particular I have shown analytically that many structural properties of horizontal-branch stars are due to the *separate* consequences of having (i) a modestly expanded helium core, and (ii) luminous flux from that core. We can understand the luminosity of horizontal-branch stars (e.g. why they are so less luminous than their red-giant-tip precursors). In addition, I obtain an analytical quenching theorem as a core-luminosity is first introduced into previously non-luminous cores: the shell-luminosity is quenched by two-thirds of the newly introduced core-luminosity. I have combined the consequences of the analysis with studies of horizontal-branch models forced to cross the Hertzsprung gap because of carefully imposed changes in their deep interiors. Such models play radius and energy catch-up with correspondingly altered quasi-static counterparts, producing a characteristic and understandable luminosity dip as they cross the HR diagram. I relate this behaviour to the similar effects seen in early post-main-sequence behaviour of intermediate-mass stars. Thus, the theory and its results, through application to such double-energy-source models as those for horizontal-branch stars, also have implications for the evolution of more massive stars across the Hertzsprung gap.

Acknowledgments

The approach employed throughout this paper owes much to Fred Hoyle's inspiration, as indeed does a good deal of the work I have done throughout my career. I owe Fred an enormous debt of gratitude, and I am honoured to be able to dedicate this paper to his memory. I thank Peter Eggleton for generously providing his versatile evolutionary code, and former student Eric Sandquist for producing Figures 10.2 and 10.3, which were obtained from its use. I am grateful to Professor L. Wilets for an informative conversation. I thank Richard Sword of the Institute of Astronomy for drafting Figures 10.1 and 10.4 from my scratchy sketches, Di Sword for translating my LaTeX latin into CUP's vulgate vernacular. I thank also the Institute's Director, Douglas Gough, for his kind invitation to contribute to this memorial tribute to Fred; above all I thank him for his much-tried patience.

Appendix 10.A   A short primer on the *UV* plane

Much has been written on the properties of stellar models in the *UV* plane, particularly in studies of red giants. A good deal of this has been unnecessarily complicated, shedding perhaps more heat than light on the topic. In order to avoid Renzini's strictures against the use of 'secret arcane methods', the use of this tool in my own approach to understanding red giants is severely limited. I employ it mainly as a scalpel for one precise and delicate purpose – the determination of the critical value for the parameter τ in the temperature--radius relationship (10.2) rather than as a gruesome bludgeon (see Appendix 10.A.2; since some introduction to the basic *UV* plane itself may be necessary, I give that in Appendix 10.A.1). However, I have found recently that several oft-repeated claims about the implications of the *UV* plane, for aspects of red giants that particularly interest us (mass storage and very large radii), are either misleading or simply wrong. Certain myths have been repeated for almost a quarter of a century. I lay those myths to rest in Appendix 10.B. I emphasize that the material in Appendix 10.B is not part of my own approach to the structure of stars with dense cores. Rather, it illustrates a long-lived misuse of the *UV* plane. It is not essential to read it to appreciate my more direct and explicit approach; however, reading it should certainly underline the advantages of the latter.

10.A.1   *Basic UV variables and relationships*

In commonly used notation, the 'homology variables' or 'homology invariants' U and V are defined by

$$U = \frac{d \ln M_r}{d \ln r} = \frac{4\pi r^3 \rho}{M_r} = 3\frac{\rho}{\bar{\rho}} \tag{10.A.1}$$

and

$$V = -\frac{d \ln p}{d \ln r} = \frac{\rho}{p}\frac{G M_r}{r} \left(= \frac{G \mu \beta M_r}{\Re T r} \right), \tag{10.A.2}$$

where the last expression for V (in parentheses) is an equivalence which holds for the usual case of ideal gas plus radiation: this will be useful later. The variables are local quantities throughout.

Logarithmic differentiation of U and V yields[12]

$$\frac{dU}{U} = 3\frac{dr}{r} + \frac{d\rho}{\rho} - \frac{dM_r}{M_r} \tag{10.A.3}$$

and

$$\frac{dV}{V} = -\frac{dr}{r} + \frac{d\rho}{\rho} + \frac{dM_r}{M_r} - \frac{dp}{p}. \tag{10.A.4}$$

Defining a purely local polytropic index, n, at any point inside the star by the local *structural* relationship

$$d\ln p = \frac{n+1}{n}d\ln\rho, \tag{10.A.5}$$

the preceding terms in ρ can be written in terms of p. After only modest manipulation, this yields

$$\frac{dV}{dU} = \frac{V}{U}\frac{U+(n+1)^{-1}V-1}{3-U-n(n+1)^{-1}V}. \tag{10.A.6}$$

(Note that one should *never* think of writing $(n+1)/n$ as γ in equation (10.A.5). That frequent practice is utterly abhorrent. It confuses a structural relationship with an isentropic one governing adiabatic changes. A clear mind will always draw this necessary distinction.)

The local polytropic-index variable n is generally determined by the equation of energy transfer, but it can also be strongly affected by molecular-mass changes, particularly if the latter are fairly abrupt. It may be constant, or essentially so, in large portions of a star. If n is constant in only a portion of the model or throughout the entire model, one has an incomplete or complete polytropic solution (a 'polytrope') respectively. With the possible exception of singular points, any zero in either the numerator or the denominator of the rhs of equation (10.A.6) defines a locus along which any solution trajectory must currently be horizontal $(dV = 0)$ or vertical $(dU = 0)$. Two such loci, necessarily straight lines, are merely the axes of the UV plane. For polytropes, the other special loci are also straight lines, connecting certain points on each of the two axes. The 'line of horizontals' runs from $(1, 0)$ on the U axis to $(0, n + 1)$ on the V axis. Correspondingly,

---

[12] Mathematicians might well cringe on seeing these bare differentials. One could rewrite all equations like these in terms of appropriate differential coefficients, $d\ln U/d\ln r$, etc. However, a better choice would be $d\ln U/d\ln s$, where s is a variable measuring arc length along a solution trajectory in the UV plane. This is because the behaviour of '$d\ln r$' (more properly $d\ln r/d\ln s$) in stars with compact cores has itself been quite badly misinterpreted; that is the subject of Appendix 10.B. Thus I invite the discerning but disconcerted reader to interpret bare differentials as such differential coefficients throughout what follows.

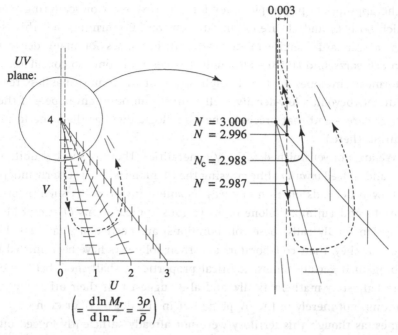

Figure 10.4 The *UV* plane for polytropic envelopes of index 3. Left-hand view: the grosser features of the *UV* plane; right-hand view: a blow-up of the small-*U* region near the *V* axis (not to scale!), where envelope integrations start for models with dense cores. Both mass-storage and the connected behaviour of envelope solution trajectories bifurcate strongly according to the starting point being above or below the line of horizontals (see text).

the 'line of verticals' runs from the centre of any model, $(3, 0)$, to $(0, 3(n + 1)/n)$. The locations and slopes of these critical lines are clearly n-dependent. Their point of intersection I_n, namely $((n - 3)/(n - 1),\ 2(n + 1)/(n - 1))$, lies outside the physically relevant quadrant to the left of the *V* axis for $n < 3$, and inside it for $n > 3$. (This leads to some differences in the possible topologies of solutions, which need not largely concern us. Despite some claims to the contrary, the 'looping behaviour' of red-giant trajectories in the *UV* plane does *not* depend critically on whether opacities yield a 'natural n' that is either greater than 3 or less than 3 outside the shell source.)

In the special case of $n = 3$, the lines of horizontals and verticals meet at I_3 on the *V* axis at $(0, 4)$. This is the case illustrated in Figure 10.4. (For simplicity, I have used mainly electron-scattering opacity throughout this paper. Whatever the initially dominant opacity source in the envelope outside the shell source, red-giant shell conditions are driven in the direction of electron-scattering-opacity dominance as M_c increases. As Hoyle and Lyttleton showed years ago, $n = 3$

is the appropriate polytropic index for extensive electron-scattering regions in which both L_r and M_r are essentially constant. Confirming that this is so has been a standard exercise in stellar-structure courses for many decades – see also Schwarzschild (1958) – although it is generally only so considered for the outermost envelopes of stars. I emphasize that the almost massless region immediately beyond a low-density shell source *is* an outer envelope – of the much denser core – and that precisely the same logic dictates the same form of the solution there.)

We are not yet finished with *UV* generalities. There is another quite remarkable and indeed *invariant* line crossing the *UV* plane which is utterly independent of how ρ depends on p in any way. Japanese investigators such as Sugimoto, Nomoto and Fujimoto, alone or in various combinations ('General Motos', as Doug Lin fondly calls these collaborations) are particularly fascinated by this line, but they, their collaborators and their followers have been misled into ascribing to it certain almost mystical properties. I shall show below how they were led astray mathematically, and also suggest that their other fondness for working, not merely in the *UV* plane but in the $(\log U, V)$ or even $(\log U, \log V)$ planes (as though this territory were not already sufficiently logged out), may also have contributed to some of the claims that do not seem to be supported by actual models.

The existence of the invariant line now under discussion may be motivated by looking back again at the differential expressions given in equations (10.A.3) and (10.A.4). They have the same density term on the rhs; therefore, subtraction will eliminate it completely, leaving a density-independent result which may be written

$$\mathrm{d}\ln V - \mathrm{d}\ln U = (2U + V - 4)\,\mathrm{d}\ln r. \qquad (10.\mathrm{A}.7)$$

From this remarkably simple equation,[13] we learn that $\mathrm{d}V/V = \mathrm{d}U/U$, or alternatively $\mathrm{d}\ln(V/U) = 0$, along an invariant line given by

$$2U + V - 4 \;(\equiv D \equiv \Delta) = 0. \qquad (10.\mathrm{A}.8)$$

(The combination $2U + V - 4$ is often denoted D or Δ, presumably denoting 'discriminant'.)

The relationship $\mathrm{d}V/V = \mathrm{d}U/U$ means that the local solution trajectory is in line with the origin. (Equivalently, $\ln(V/U)$ is stationary; it can also be shown that in passing across the line $\Delta = 0$ the trajectory is always turning to the left

---

[13] Simple though this equation is, a number of well-known authors have seriously misinterpreted its implications by writing it in another form and then drawing quite inappropriate conclusions as a result. How this sorry story came about is discussed in Appendix 10.B.

as r increases.) In other words, although the polytropic lines of horizontals and verticals rotate in opposite directions about their respective fulcrums ($U = 1$ and 3) on the U axis as a function of n, *any* solution trajectory whatsoever points directly towards or away from the origin when it crosses the invariant line given by equation (10.A.8). There seems to be no standard name for this line, but I call it the 'line of tangents' (tangent to the origin, understood). That line (which is also the locus of singularities I_n where the lines of horizontals and verticals cross for any n) is also shown in the figure. The knowledge that non-singular solutions always cross it in line with the origin, whatever the local structure, helps one to visualize the general shape of solution curves. The rest of the general topology may now be appreciated, bearing in mind that the slope of dV/dU changes sign every time a line of verticals or horizontals is crossed.

In the next section I shall show that stars with low-density shells developing outside dense cores, but also possessing reasonably massive envelopes, must necessarily have a UV-plane trajectory that crosses the line of tangents from larger V to smaller V at small U, close to the V axis. Consequently, given the negative sign of Δ below the line of horizontals, the model trajectory must pass continuously to yet lower values of V/U (i.e. generally down to lower V and to the right to larger U) as r increases, before finally re-crossing the line of tangents into the upper region once more. There, Δ is positive and the general sense of progress is reversed back to higher V/U values. This produces the characteristic 'looping' behaviour of giant models in the UV plane. The general reason for such behaviour lies in the necessity of storing substantial mass outside a dense core; without 'looping', very little mass can be stored, as will be shown. Despite the implications of both Yahil and van den Horn (1985) and Applegate (1988), the stellar-structure limitations that result in stars becoming red giants do *not* depend on whether the 'natural n' defined outside the shell region of a protogiant is greater than or less than 3.

10.A.2   *Mass storage and the critical value of* τ

Summary of what will be shown, and a useful new variable, N

The critical nature of the value for τ follows from the remarkable differences in envelope mass storage that become possible as a function of very small changes in the starting or basal envelope temperature, T_{sh}, for a given value of $\rho_{sh}/\bar{\rho}_c \ll 1$. This extreme sensitivity implies the following behaviour in the UV plane: over a very small range in basal temperature the resulting envelope trajectories change from (i) going essentially straight up along the V axis, to (ii) dramatically looping down in V and out to substantial values of U. Cases of type (i) store very little mass, while those of type (ii) can store large, even HUGE amounts of mass. I shall first show this for a purely polytropic envelope with $n = 3$; but with appropriate changes, similar results hold for other polytropes or for more realistic radiative envelopes. (Extension to envelopes convective in their exteriors follows that.)

It is not easy to draw just one simple figure that can simultaneously illustrate these wildly different behaviours and show why they must necessarily occur. Partly, this is because very low values of $\rho/\bar{\rho}_c$, and therefore of U, are known to arise in the vicinity of the shell and in the surrounding radiative zone. (Remarkably, no-one else seems to have demonstrated explicitly the *physical necessity* of such low values previously, as I have above, as a simple consequence of equation (10.8), for example; it is yet another solution property commonly taken to have been proved by familiarity.) Because of this awkward feature of contrasting numerical scales – extremely small U co-existing with other values reaching of order unity – many investigators have opted for the use of the $(\log U, V)$ plane, which apparently stretches the left-hand part of the UV territory into extended and seemingly boundless plains, or even the $(\log U, \log V)$ plane which stretches out both the small-U and small-V regions of the diagram. However, there are serious drawbacks to this apparent ingenuity: the beautiful geometrical simplicity and implications of all the critical straight lines for polytropes, as discussed above, are then lost. In my view, that is far too heavy a price to pay. Indeed, I shall show that certain repeated misconceptions may have their origin partly in the latter overstretching act.

I prefer instead to show a finite part of the plane for an overall view, and also to blow up the critical region in which U_{sh} $(= 3\rho_{sh}/\bar{\rho}_c)$ and V_{sh} both reside. (In that way, I can still use the remarkable implications of all the critical lines directly.) Those two views are sketched in Figure 10.4, for the case $n = 3$. It is still difficult to show every desirable feature to scale (and no less so in the alternative versions of the UV plane), which is why this is just a sketch. However, I shall quote some explicit values that help to make the necessary points.

Also shown in the blow-up is another convenient new variable, N, which helps parametrize the starting temperature for envelope integrations, T_{sh}. Recall that the leading term in the temperature formula (10.2) has a coefficient of $1/4$. The 4 is there because, in effect, n is locally 3; for general n, the coefficient would be $1/(n+1)$. Since I am going to show that interesting values of τ are generally small, in appropriate relative or dimensionless terms, it is convenient to introduce this new N (whose value will prove to be $<$ but $\simeq n$) by

$$T_{sh} = \frac{1}{N+1} \frac{G\mu\beta M_c}{\Re} \frac{1}{r_{sh}}. \tag{10.A.9}$$

Notice that, by intent, there is no τ in this equation. Through the factor $N + 1$

the new variable N has in a sense taken up the previous role of τ in providing for possible variations in the value of T_{sh}. In fact, $N + 1$ is none other than the value of V_{sh}, as can be seen from the alternative definition of V in equation (10.A.2). I write T_{sh} and V_{sh} in this form because the following results, illustrated explicitly for $n = 3$, are also generally true for all values of n approaching 3 from below. (A slight variant holds for $n > 3$.)

Here then is my crucial conclusion, which I shall proceed to demonstrate and derive in the next subsection: mass storage is relatively modest for $N > N_{crit}$, and much more substantial for $N < N_{crit}$, where

$$N_{crit} = n - (n + 1)U_{sh} = n - 3(n + 1)(\rho_{sh}/\bar{\rho}_c). \tag{10.A.10}$$

What is more, as $n \to 3$ from below, 'relatively modest' becomes 'negligible', and 'more substantial' becomes 'huge', once one is dealing with very small values of $\rho_{sh}/\bar{\rho}_c$. Note that this critical value for N corresponds to having the starting point for envelope integrations lie precisely on the line of horizontals.

This is the major result to be proved in this appendix, so I restate it in its physical context for further emphasis. For those opacities that yield a 'natural n' moderately less than or equal to 3, there is a watershed for envelope mass storage given by having a precisely determined functional relationship for the envelope's base temperature. That base temperature is such that it places the starting point for envelope integrations in the UV plane precisely on the line of horizontals. All envelopes with starting points either there or above it store very limited amounts of mass; to store substantial mass (comparable with or exceeding a reasonable fraction of the core mass), the envelope must start below the line of horizontals, but only minutely so because of the general topology there for small U_{sh}. That is the essential mathematical condition. Corresponding to it, and to a close approximation in practice for all such n, the leading term in the general temperature formula (10.2) is supplemented by a critical value for τ given to first order by $\tau >$ but $\simeq 3(\rho_{sh}/\bar{\rho}_c)$.

Graphical reasons for the behaviour cited

The blown-up portion of Figure 10.4 illustrates the nature of the topology of the UV plane for regions both with very small U and with V close to 4 (i.e. N near 3), for the case $n = 3$. Taking $\rho_{sh}/\bar{\rho}_c = 10^{-3}$, to define a starting value for ρ_{sh}, we have $U_{sh} = 0.003$ as the starting value of U for the outer envelope. So, any $n = 3$ envelope's mass-storage capability for this particular starting density is completely determined by the value of N, where $N + 1$ effectively acts as a dimensionless inverse surrogate for T_{sh}.

I shall not give an extensive tabulation of results here. Those for N around a critical value 2.988 are particularly striking, and illustrate the main point in themselves. Letting M_* and R_* be respectively the mass and radius of the entire model (core plus envelope), I find that

$$N = 2.989 \Rightarrow M_*/M_c = 1.020, \qquad R_*/R_c > 4.35 \times 10^3,$$

whereas

$$N = 2.987 \Rightarrow M_*/M_c > 33.26, \qquad R_*/R_c > 5.53 \times 10^5.$$

These, then, are the key results: a change in $N + 1$ and therefore in T_{sh} of only 1 part in 4000 on either side of a critical value is sufficient to change the mass storage in the $n = 3$ envelope from only 2% of M_c to over 32 times M_c, a change in mass-storage capability by a factor of more than 1600. And all this from changes in the initial variable of only $\pm 1/4000(!)$. In this incredibly small range of $N + 1$, or basal temperature, mass storage verily goes from famine to feast. Meanwhile, the additional radial extent, already nothing to be sneezed at for $N = 2.989$, itself increases by more than two orders of magnitude over this small range in N. As these results suggest, the critical value of N for mass storage (and incidentally for marked changes in radial extent also) is in fact 2.988.

How may we understand this initially quite astonishing effect on potential mass storage? This is the one place where I shall actually use the topology of the UV plane to help promote insight. Fortunately, it involves looking only at the region where the various critical lines converge. Consider the starting point for envelope integrations in Figure 10.4. U_{init} is 0.003, while N characterizes position along that starting line-up for U. Four characteristic or specific starting points are shown in the diagram. High enough for any starting point above the line of verticals ($N > 2.996$) the external solution slopes backwards quite steeply towards the upward vertical V of the UV plane. Clearly the value of U ($\equiv d \ln M / d \ln r$) can never amount to a hill of beans, to use that colourful phrase of our transatlantic cousins. Thus Martin Schwarzschild's and Fred Hoyle's assumptions (reached by different routes) that $N + 1$ (in our language) was 4 (or $N = 3$) was too conservative, if only slightly so, and insufficient for mass storage. (Forgive me, Fred, for accusing you of being conservative – if only here.) Beginning *on* the line of verticals ($N = 2.996$), even though the trajectory starts straight upwards, is not good enough; the trajectory is quickly forced to join the same kind of flow as the others with small mass storage. Furthermore, starting on the line of horizontals (at $N = 2.988$), the initially horizontal direction of the solution trajectory, towards larger values of U, might at first sight look promising, but alas, it too is inevitably swept upwards to cross the line of verticals before it has gone very far, resulting yet again in rather small mass storage.

Thus, in the manner of a classical *reductio ad absurdum* argument, the only possibility left for significant mass storage is to start with $N < 2.988$. And *how* the model takes advantage of this one remaining possibility! While $(dV/dU)_{sh} = 0$ for $N = 2.988$, its value is astonishingly large and negative, $\simeq -49.2$, for $N = 2.987$. Thus, the solution trajectory initially heads downwards extremely steeply. (The slope for $N = 2.989$ has almost the same magnitude, but with positive sign of course.) In contrast to all the trajectories for $N \geq 2.988$, which are trapped into flows severely limiting mass storage, that for $N = 2.987$ and similar values less than 2.988 can head down for the relatively open territory of smaller V and much larger U, ultimately greater or equal to unity, where significant mass storage may now, and indeed does, take place. As indicated above, a simply huge change in mass-storage potential takes place for those trajectories lying initially between $N = 2.988$ and 2.987. That change is so large that the more practical requirement – that an $n = 3$ envelope should store only of order $0.5–4.0 \times M_c$ (to cover the range required on the giant branch, from evolutionarily inverted top to bottom respectively) – tunes the temperature relationship (10.2) extremely precisely. Extremely large radii necessarily follow as part of this very fine tuning – but always as a consequence of the prime directive (if one unacknowledged in significance and even derided by Renzini), namely to find a way to store reasonable envelope masses. Several further points need to be made in this regard:

(a) Polytropic envelopes with $n = 3.25$ (a more appropriate approximation deep down for Kramers-opacity cases) behave rather similarly, except that now there is a 'cut' in the UV plane which, for $n > 3$, plays the part formerly played by the line of horizontals for $n \leq 3$, in dividing envelopes with negligible mass storage from those with substantial mass storage. Yahil and van den Horn (1985) went so far as to suggest that a 'cut' of this nature was the essential ingredient for giganterithrotropism (a coinage abetted by Eggleton); but as we have seen above, that is far from the whole story.

(b) Fully computed radiative envelopes behave like $n = 3$, or (*mutatis mutandis*) $n = 3.25$ polytropic envelopes. Thus are explained the results of Sandage and Schwarzschild (1952), who found that radiative envelopes (with no allowance for the possibility of a convective attic) swelled up and became quite grossly large once dense cores in excess of the Schönberg–Chandrasekhar limit were encountered.

(c) There is another, more physico-mathematical way to appreciate the general large-looping shape of the UV-plane trajectories when storing substantial mass (in the sense of many times M_c) in envelopes, and the reduction of this loop size as a smaller multiple of M_c is stored. First consider complete polytropes of index $n = 3$ or 3.25. These go all the way with a smoothly changing negative slope to the model's centre at the point $U = 3$ on the U axis. Now place a small dense central pea of matter at the model's centre. By topological continuity, only at the last moment, going inwards, will the polytropic solution – now an incomplete (*envelope*) solution – veer away from the vicinity of $U = 3$ and travel around on a very long, low-V and finally low-U, axes-hugging trajectory to its ultimate destination close to the V axis near $V = 4$ or 4.25 respectively. Such a trajectory corresponds to an extreme example of what older workers called a 'condensed' solution, which has $r_c = 0$ at a finite M_c. (This terminology was always a little confusing, not unlike the use of the term 'missing matter', when of course it is the *light* that is missing from matter otherwise deduced to be there. In the present case, it is not the envelope that is condensed, but rather the underlying core, necessarily *not* part of the envelope solution. Similar remarks apply to the even more confusing old term 'collapsed solution', for an envelope solution somehow *not* collapsing even though the space beneath it is devoid of matter: $M_c = 0$ at a finite r_c. Such 'collapsed' solution trajectories pass above the point $U = 3$ on the U axis and proceed to an unphysical and aetherial limit at ever larger values of U.)

Clearly, the greater the mass fraction stored in the dense central part, the smaller the loop in the UV plane must be, until such time that the envelope stores hardly any mass and its trajectory then runs close to the upper V axis for reasons already given. Earlier workers always used to refer to 'the great convergence of inward integrations' for incomplete polytropes, towards the point $V = n + 1$ on the V axis. (Schwarzschild (1958) makes such a reference.[14]) The converse of this viewpoint is far more important for us, physically, and is indeed precisely what I have been stressing above: we can now understand the

---

[14] Thus he effectively took $\tau = 0$ in my equation (10.2). Hoyle did the same, if for different and more physical reasons. He argued that τ could not be large and negative; that would simply give a classical, essentially massless envelope. If l_T were the temperature scale height, the ratio l_T/r would tend rapidly to zero. On the other hand, τ large and positive would lead to the opposite result, with l_T/r tending to ever larger values and an envelope of almost infinite extent and infinite mass-storing capability, far more than was needed. So he argued, almost intuitively, that τ must be small in absolute terms, and therefore it should be good enough to take it to be zero. Not quite!

extremely fine tuning of the temperature relationship (10.2) as the corresponding property when observed from the opposite point of view, namely the necessity of storing sensible, desired core-mass multiples in outward integrations outside dense cores.

(d)   Physically, mass-storage possibilities are not confined to those of purely radiative structures. Suppose that the solution is proceeding outwards along its radiative path in the $(\log \rho, \log T)$ plane, i.e. as something like $\rho \propto T^3$ initially. Correspondingly, the mass stored (relative to M_c) grows only very slowly initially – first logarithmically and then algebraically, but always with a low relative density multiplier, $3\rho_{sh}/\bar{\rho}_c$. If, at some point, the solution switches to a convective structure $(\rho \propto T^{1.5})$, additional mass may now be stored with increasingly greater efficiency as T continues to drop. Such a switch not only has the great merit of satisfying mass storage requirements in a significantly smaller radius than that which we have shown to be needed in the fully radiative case; it also enables the surface boundary condition to be satisfied too. In fact, a completed model must do both.

   In his famous paper with Schwarzschild, Fred Hoyle stressed the latter necessity, with the former aspect somewhat subsidiary. In his later lectures to and his talks with his students, however, he came to see the necessity of solving the mass-storage problem as the key element setting the scale for temperature in the deep interior as a function of other conditions there (as I have tried to show), with the surface boundary condition as another but slightly subservient external requirement. It was then, and is now, absolutely clear to me that while both conditions were essential, the necessity of arranging for substantial mass storage had definitely gained primacy in Fred's mind. This may seem a rather pedantic difference to some readers, but I have tried to show in this paper that it is not only (i) the *inability* to store mass close to a dense core but also (ii) the *preservation* of the ability to store it much further out. Together these provide the key temperature relationship (10.2), from which all else follows. What I like to call the 'expansion imperative' follows from these conditions deep down, all determined by the radiative regions immediately surrounding the core, *whatever the detailed structure of the rest of the outer parts of the star*. Thus, mass-storage difficulties are the prime and ultimate strategic problem for the star as a dense core develops. To them alone may be traced the tendency for post-main-sequence stars to swell into red giants. All else is secondary, a question of the tactics employed to meet the strategic imperative.

(e)   Since there is some persistent confusion in popular or semi-popular circles (note the author's tact in not mentioning certain professional authors), let me emphasize that convective envelopes make red giants *smaller* than they would otherwise have been. Thus, red giants most definitely are not large(r), but rather small(er) because they have convective envelopes rather than radiative ones.

(f)   Finally, we observe that just as substantial mass storage in a fully radiative envelope requires that such an envelope start near $V = 4$ or 4.25 (minus some small correction) for small $\rho_{sh}/\bar{\rho}_c$, so, analogously, efficient storage in the ultimate convective superstructure with $\rho_{base}/\bar{\rho}_{base}$ still small requires that, at its own base, V be near 2.5 (minus some small correction). This could *never* be possible unless the solution in the radiative regions were able to proceed downwards in the UV plane to such a convective-envelope starting point. This confirms yet again the necessity of first passing into the region below the fully radiative generalization of the line of horizontals. In this final, ultimate solution of the dense stellar core's external mass-storage problem, there is a subtle interplay between the two separate 'τ' values (τ_{rad} and τ_{conv}) now required in the two distinct temperature relationships (for the separate radiative and convective envelope regions) which enable both the mass storage and the outer boundary conditions to be met simultaneously. The previously extremely tight limit on τ_{rad} is slightly relaxed, but the steepness of the UV-plane trajectory for small U still ensures that its coefficient in the relation (10.3) remains fairly close to 3.

I have now completed my task of demonstrating how the critical value of the constant τ in the temperature relation (10.3) depends upon the value of $\rho_{sh}/\bar{\rho}_c$. I consider that to be the sole reason for needing and therefore introducing the UV plane. In my opinion, that is all – and perhaps more – than the average reader needs (or can take). The following appendix is addressed to the more expert reader who may be aware of other claims made on behalf of the UV plane, but has never had the stomach for examining them carefully. It is not for the UV faint-of-heart. *Caveat lector.*

Appendix 10.B   Red giants and 'that gruesome tool: the UV plane'! - some remarkably repeated confusion

When refereeing my first paper on the nature of the horizontal branch, Martin Schwarzschild expressed some justified reservations about the validity of arguments I had given to help explain or understand my computed results. He went on to assert that he felt that only the Japanese workers were ever likely to understand such results, 'by means of that gruesome tool: the *UV* plane'! Although Schwarzschild himself had used the *UV* plane with consummate skill in his model computations, he still detested it as an analytical tool, yet admired its apparently expert use in the hands of Hayashi and his school. Schwarzschild's own instincts in stellar structure were generally so soundly based that I too adopted his deference to Hayashi's school in all matters to do with the *UV* plane. It has therefore come as something of a shock recently to learn that that deference and implicit trust were misplaced where previous *UV*-plane 'explanations' of significant red-giant and horizontal-branch star properties were concerned.

I should have realized that something was seriously wrong with certain claims in this area immediately following a rather odd occurrence at the 1996 meeting (the 'Ickofest') on the Isle of Elba celebrating Icko Iben's 65th birthday. I had given a talk about the importance of the envelope mass-storage problem as a key to understanding the structure of low-mass red giants in particular (i.e. low-mass hydrogen-shell-source stars with dense, essentially zero-luminosity, degenerate helium cores), although at the time I had not yet discovered either of my two main asymptotic theorems: equations (10.7 and 10.8).[15] Following my talk at the Ickofest, Sugimoto leaped forward to give a highly critical response in which he said that I was completely wrong in thinking that mass storage presented any particular problem or could yield any insight about the structure of red giants. (On reflection, this was rather an odd objection since scattered references to such a problem and its consequences, but in the context of his own *UV*-plane considerations, occur throughout his own papers cited in this section.) He then proceeded to tell us all that the entire key to red-giant structure lay in something he felt I had not stressed sufficiently, although it was certainly implied as part of my considerations: the necessity for the *UV*-plane solution trajectory crossing the line $D = 0$ (i.e. my 'line of tangents') twice. He argued that once the first crossing had occurred (at small U, its smallness taken for granted), that fact in itself necessarily implied large looping behaviour in the *UV* plane.[16] Furthermore, Sugimoto said that the externally directed increments in $d \ln r$ would then necessarily become very large indeed, as a result of these crossings and/or the implied proximity of the resulting *UV*-plane trajectory to

---

[15] My post-meeting discovery of the first asymptotic theorem was announced in an abstract that appeared in the meeting's proceedings (Faulkner 1997).

[16] Careful readers of Appendix 10.A.2 will have realized that this is not so. Crossing the line of tangents is simply not sufficient; as already shown, the trajectory must also have crossed the line of horizontals. I shall return to this point after discussing the main mathematical misconception.

the line $D = 0$. He made several references to this having been exhaustively and conclusively demonstrated by himself and Nomoto (Sugimoto and Nomoto 1980).

I confess that although I understood, and already appreciated, some of the things that Sugimoto said, others struck me as being rather misplaced, putting far too much emphasis on the existence of the line $D = 0$ and either the crossing of the line itself or a presumed proximity to it. I thought I might have misunderstood his point, or his language in person, but his published account (Sugimoto 1997, Section 3, p. 21) contains the passage: '... in the red giant region of the $U - V$ plane, the envelope solution runs close to the line of small D so that the integral of equation (3.3) i.e., the radius change is very large.' (His equation (3.3) is the very one, my equation (10.B.1) below, whose misuse I discuss in this appendix.) Sugimoto's verbal misrepresentation of the envelope trajectories and the badly misinterpreted equation to which he refers involve a number of misconceptions which I have since found to run through several successive papers in this context, from the original source (Sugimoto and Nomoto 1980) through Fujimoto and Iben (1991), Dorman (1992), Sugimoto's referenced 1997 paper, and Sugimoto and Fujimoto (2000). (My confidence in Sugimoto's published 1997 assertions was not increased by reading his Section 1: a very strange description of my Elba talk in which he praised a mere afterthought as though it were my central thrust, then introduced my main topic – classical low-mass red-giant structure – as though it were his own original addition to the discussion; and Section 2: an unconvincing account of the supposed consequences of what he called a 'virtual gravitational contraction of the core', containing unsubstantiated assumptions and assertions about implications he claimed were obviously very different in two separate cases, for reasons never made clear.)

An elementary misinterpretation of a rewritten form of equation (10.A.7) appears to be at the root of some specific problems running through all of the papers just cited. That rewritten form is as follows:

$$\mathrm{d}\ln r = \frac{\mathrm{d}\ln V - \mathrm{d}\ln U}{2U + V - 4}. \tag{10.B.1}$$

Most of the authors cited, among them Dorman (1992), similarly write:[17]

$$\mathrm{d}\ln M = \frac{U(\mathrm{d}\ln V - \mathrm{d}\ln U)}{2U + V - 4}. \tag{10.B.2}$$

I shall now continue by quoting Dorman in particular, as a native (albeit transatlantic) English speaker, since in his case at least, inapt verbal characterizations

---

[17] Some authors write ln (natural logarithm, to base e) and others log (those to base 10) in their equations, while others use ln in the text but plot log in their diagrams. The difference involves inessential scale changes, the form of the equations remaining otherwise unaltered, of course.

cannot presumably be attributed merely to language difficulties. In fact, his descriptions of the mathematical situation and its implications are essentially no different from what the other authors write, assume or imply.

Immediately after exhibiting the previous two equations (10.B.1) and (10.B.2) (his equations (A20) and (A21)), Dorman writes:

> The left-hand sides of these expressions diverge[18] on the line of singular points (A19). If a solution trajectory lies close to this line, therefore, its radius will tend to be large compared to a solution which stays in the upper portion of the (U, V) plane... This provides an elegant conclusion to our mathematical summary which unites two separate topics within the theoretical framework.

I do not find this to be an elegant conclusion at all, since it starts from a demonstrably false premise. It is *simply untrue* that the value of the expressions on the right-hand sides[19] of the above equations diverge *on any red-giant or horizontal-branch solution, or indeed on any physically possible model trajectory in the UV plane, however realistic or idealized* – and that is surely the point at issue. We need to reject firmly Dorman's descriptively deceptive language 'line of singular points', and similarly Fujimoto and Iben's 'critical curve'. These terms are extremely misleading, and only show how deeply embedded and seductive is the notion that solutions necessarily behave pathologically in the vicinity of the line of tangents.

Both the mathematics and the language used to describe its consequences seem to be extremely confused in this succession of papers. For example, one author writes: 'runs... near the line of $V = 4$' and 'runs close to the line of small D' (in the small-U region – my clarification) when the context makes it clear that he means 'approaches' or 'crosses' in both instances. Others write: 'must run along the critical curve' when they really mean: 'must make a loop'. Broadly speaking, then, in these papers the great extent of red-giant radii is attributed to either (i) the smallness of $2U + V - 4$ (i.e. D or Δ) as the line of tangents $D = 0$ is approached and/or crossed, or (ii) the proximity of a solution curve to that line, as measured by the small value of D over a substantial range in U/V. Several issues thus need to be addressed, according to whether (i) 'crossing' or 'approaching' or (ii) 'lying close to', 'running near' or 'running along' the line of tangents is meant.

---

[18] Sic. Already we have a problem. Having written 'left-hand sides' he must also mean 'of these equations', and not 'of these expressions'. But it seems most probable that he means to convey, instead, 'the values of the expressions on the *right-hand sides* of these equations diverge and therefore the left-hand sides must also diverge...'. I shall proceed as though that accurate description of an incorrect argument is the case.

[19] See preceding footnote.

Before I do so, consider for completeness other possible uses of the same line of reasoning which has concluded that there is somehow singular or divergent behaviour in the proximity of the line of tangents. The lines of horizontals or verticals correspond to $d\ln V/d\ln r$ or $d\ln U/d\ln r = 0$, respectively. Inverting either of these equations would similarly 'imply' a divergence, using the same logic. The same false conclusion would hold for $d\ln r$ with respect to any variable containing a non-tangential component at any arbitrary point. Thus, one could attribute extremely large radii to the solution trajectory having a turning point anywhere one chose. (That is why I remarked that $d\ln r/ds$ is a much better measure.)

10.B.1   The radial extents

First, 'crossing the line'. The appropriate starting point for this or any related analysis is the equation (10.A.7) I derived above. This shows that on the line $2U + V - 4 = 0$, unless one has some additional reason to believe or know that $d\ln r$ behaves singularly (an unphysical impossibilility, in fact) the simplest and most straightforward implication – the one that holds, in practice – is that $d\ln V - d\ln U = 0$, no more and no less. (Physically, this result is mandated by the following consideration. The only way that $d\ln r$ could be singular would be if r were able to be literally zero on the outer edge of a region of finite mass.) Therefore, the only rational implication, as stated following my equation (10.A.8), is that *all* solution trajectories that cross the line of tangents do so in such a manner that the origin of the UV plane lies on the tangent at that point of crossing – hence the name I give it. In other words, all solutions 'point at or away from' the origin as they cross the line of tangents. (There is nothing singular, and nothing divergent, about that. In fact it is a considerable help in appreciating the general topology of the UV plane in such cases.) Equivalently, all solutions in the $(\ln U, \ln V)$ plane cross the curve corresponding to $D = 0$ at 45 degrees – a property mentioned frequently by the Japanese workers in particular.

So, the value of $d\ln r$ (more properly, $d\ln r/d\ln s$; see footnote referring to equation (10.A.3) remains perfectly finite as any solution trajectory crosses the line of tangents. That is why, in equation (10.A.7), the zero in $2U + V - 4$ implies that $(d\ln V - d\ln U)$ is correspondingly zero there. Furthermore, whatever value $d\ln r$ has at that point, it is itself varying smoothly along the trajectory, as the two parenthetical expressions just mentioned approach their simultaneous zeros with the same order of smallness. The situation is just like any other in which a limiting argument is used to determine the *finite* value of an expression at a point where both numerator and denominator tend to zero simultaneously (e.g. l'Hôpital's rule). Put frankly, it simply makes no sense to fasten one's attention on just $2U + V - 4$ becoming zero in equation (10.B.1), and then to conclude (falsely) that $d \ln r$ somehow behaves badly, even divergently there. The 'oddity' in $2U + V - 4$ becoming zero has been 'used up' in making $(d \ln V - d \ln U)$ zero also. One cannot ignore the already-established simultaneous zero in $(d \ln V - d \ln U)$. This is like trying to have one's mathematical cake, and eat it too.

Sugimoto and Nomoto (1980) were apparently the first to make this misinterpretation. In a section entitled 'Reason why the star becomes a red giant', they write:

> When the core becomes of condensed type...the $U - V$ curve makes a loop, i.e., crosses $\Delta = 0$, and runs in the region of small value of U with negative Δ...When N is close to 3, the absolute value of Δ is thus very small, which makes the stellar radius very large according to Equation (2.4).

Their equation (2.4) is my equation (10.B.1). Thus, this passage explicitly attributes large radii to having $-\Delta$ very small in the region of small U. This makes it a 'crossing' rather than a 'running along' explanation. Were this reasoning correct, it ought to hold for both polytropic envelopes surrounding dense cores (whether those envelopes store negligible or large amounts of mass) and for more realistic, mass-storing giant envelopes that satisfy the same conditions. It simply does not, as some examples will shortly show. Large envelope radii cannot be attributed directly to having small $-\Delta$ in the region of small U per se. If one wishes to find a 'cause' for large radii in the UV plane, it must be sought in other aspects of the topology far from the first crossing of the line of tangents, in regions ultimately reached only because of a quite different 'cause' at small U.

Three examples should suffice. They will help illustrate not only the 'crossing the line' fallacy, but also the deficiency in the 'running along the line' claims. The first two examples involve polytropic envelopes described earlier, the third, more realistic models of red giants:

(a)   A polytropic envelope with $n = 3$ starting below the line of tangents but just *above* the line of horizontals, at $(U, V) = (0.003, 3.989)$, ultimately has a radius of ~ 4300 times the core radius. But, of this extension by a factor of ~ 4300, only a factor of ~ 6.3 occurs by the time the solution trajectory crosses the line of tangents, and a factor ~ 8.4 as it crosses the line of verticals. Thus, by far the greatest extent in radius occurs far from the line of tangents. The reason why what appear to be huge values of $d \ln r$ (wrt either $d \ln(V/U)$ or $d \ln U$) in the original region have such little integrated effect is because the *range* covered by the solution's independent variables in that region is so minute – at its maximum, when crossing the line of verticals, U has reached only $\sim 0.003\,022$. This reminds us that a huge differential coefficient has little effect if confined topologically to a very small range of the independent variable. The reader will be sorry to learn that I have yet more moral lessons to be drawn from this example, later. (This particular model envelope stores negligible mass, ~ 0.02 of its core-mass, because U always remains very small.)

(b) In contrast, a corresponding $n = 3$ polytropic envelope starting just *below* the line of horizontals (and therefore a bit further away from the line of tangents), at $(U,V) = (0.003, 3.987)$, takes a very different trajectory, with grossly different results. It achieves a radial extension by a factor exceeding 5.5×10^5. This factor is impressive, but the bulk of it occurs far from either region of small Δ, whether at small U or at the much larger value for U (~ 1.03) where the line of tangents is crossed for the second time. Only 9% of the full radial extent is reached by the point where the trajectory crosses the line of tangents, and just 14% by the line of verticals. Thus 86% of the radial extent occurs beyond the line of verticals and therefore well away from crossing the so-called 'critical line', in the upper part of the UV plane on the way to the surface regions. (In no sense can the solution trajectory be said to be running 'near' or 'along' the line of tangents at any place whatsoever. That alternative 'reason' for large extents isn't satisfactory either.) I defer discussion of where the mass integral finds its greatest increments, except to say that neither in this case nor the next one is it where Sugimoto and Nomoto assert.

(c) Finally, in testing the radial-extent assertions of these various authors in more realistic models, I check what they say against more realistic red-giant models. Some aspects have been checked against the UV planes of models computed with the Eggleton code. Other aspects can be checked against a readily available model tabulated long ago in the literature: Schwarzschild's fairly advanced red-giant model of 1.3 solar masses, Table 28.7, p. 260 of Schwarzschild (1958). Hydrogen is exhausted in the inner 26% of this model's mass ($M_c = 0.34\,M_\odot$). Unfortunately, the UV plane for Schwarzschild's tabulation is not available, but the general features of his model agree with those of mine, and the UV-plane properties presumably do too.

The additional radial extents beyond the cores, while still large in these models, are much more modest than those in the previous two examples because of the outer convective ($n = 1.5$) envelopes. Nevertheless, they enable us to illustrate another important point. Although $d \ln r$ relative to differentials of any one of a number of possible other independent variables (s, $\ln U$, $\ln(V/U)$, ...) may tend to become very large for $U_{sh} \ll 1$, one should neither forget nor ignore the context. The cores themselves are *small*. Therefore, much of that apparently large contributor to the overall integral for additional radial extent is used in bringing the radius up to something like 'normal' (main-sequence) values. Thus, in the radiative part of Schwarzschild's model, a radial extent beyond the core by a relatively large factor of $\sim$60 still brings the radius up to only $\sim$1.7 $R_\odot$. Beyond that, in the convective region, there is then an additional, smaller *factor* of only $\sim$12.5. Nevertheless, that factor contributes the bulk of the 'large radius' of $\sim$21 $R_\odot$. Examination of the corresponding UV-plane trajectories for my own models shows no particular tendency for increments in $d \ln r$ or fractional radius to cluster around the line of tangents in any particularly pronounced way. Furthermore, the solution trajectories do not 'run along' the line of tangents in any way, and language used should not convey the impression that they do so. If one has opened up the small-U and small-V territory by using logarithmic variables, it may well appear in such diagrams that the solution remains remarkably close to the critical curve, given the enlarged area of such diagrams. But that is an illusion created by that choice of logarithmic variables for illustrative purposes.

As one final example of forgetting the context of a 'large' (or even a supposedly large) value for $d \ln r$, we have a statement that 'Both runs along the loop cause an augmentation in the radial distance, which is greater, the closer the structure curve is to the critical curve' (Fujimoto and Iben 1991). But in any giant-like star with a dense core, that outward run above the curve, never very close to it anyway, is taking the radius only out to the *small* value it has on the outside of that dense core. *It can in no way contribute to large radii outside that core.* The context has again been completely ignored.

10.B.2   *The mass extents*

We now come to the question of the additional mass extents. I can be briefer. After referencing their equivalent of my equation (10.B.2), most authors write something like: 'When the structure curve comes close to the critical curve, therefore, the mass also increases along it. In contrast with the increment in the radius, the mass increment is shifted to that part of the structure curve where U is larger' (Fujimoto and Iben 1991). The second sentence is true but hardly a revelation. However, the first once again suggests proximity to the line of tangents as the major factor. This is as incorrect for mass extents as it is for radial extents. I shall give explicit counterexamples to both of these claims before summarizing my own point of view, below. But first I need to deal with even more specific – and misleading – assertions in the paper in which this whole sorry saga originated: Sugimoto and Nomoto (1980).

Considering accommodation of the envelope mass, Sugimoto and Nomoto write:

> It can be done if the value of Δ is negative in the envelope just above the core edge. Then the $U - V$ curve can run to the direction of increasing U/V ... It takes a local maximum where the value of U/V becomes *of the order of mass fraction contained in the envelope* and then the sign of Δ changes back to positive, i.e., the $U - V$ curve makes a loop. Such structure corresponds to the envelope of a red giant star, in which *the bulk of the envelope mass is contained in outer shells around the local maximum of U/V.*' [Emphasis added.]

I take issue with both parts of this quotation that are emphasized above. The notion that the local maximum of U/V (i.e. its value as a solution trajectory crosses the line of tangents) is of the order of the mass fraction in the envelope is a very loose one. The authors give no formal justification for it. It is off in practice by factors as much as 2 or 3 for red giants, and of order 10 for those horizontal-branch models possessing loops (Dorman 1992). It is hardly a useful guide, principle, or rule of thumb. Thus, it is necessarily off for any model with a large loop in which the maximum value of U/V exceeds 1, such as the second $n = 3$ polytrope discussed above. For that, the trajectory crosses the line of tangents close to the point $(U, V) = (1.5, 1)$; the mass fraction contained in the envelope for that is indeed large, ~ 0.97, but it can hardly be ~ 1.5. In one of my own more realistic models with total mass $1\,M_\odot$, core-mass $\simeq 0.4\,M_\odot$ (i.e. envelope mass fraction $\simeq 0.6$), the crossing point is at $(U, V) \simeq (0.8, 2.4)$, giving a maximum value for U/V of ~ 0.33. It fails exceptionally badly for a horizontal-branch loop model by Dorman (1992), where the envelope mass fraction is ~ 0.44, but the crossing point (at $\sim (0.2, 3.6)$) gives $U/V \simeq 0.056$.

As for the second generalized claim, that the bulk of the envelope mass is contained in outer shells around the local maximum of U/V, I find that simply misleading, also. In the polytropic model, on crossing the successive lines of horizontals, tangents, verticals, and on reaching the surface, the radial mass variable has the following multiples of core-mass: ~1.6, 2.9, 6.4 and 33.3 respectively. By far the greater part, ~81% of the total mass, is stored beyond the line of verticals, well after the line of tangents is crossed. Once again, the situation is less extreme in the case of the more realistic models because of the effect of the reduced envelope polytropic index of 1.5. Nevertheless, I and my students have long marked points at equal intervals of either 0.05 or 0.10 in mass fraction on our own UV plots. These cluster most strongly exactly where one would expect – around the point of maximum U (i.e. near the crossing of the line of verticals), and not around the points where the line of tangents is crossed. Once again, this is because there is nothing odd about the behaviour of $d \ln r/ds$ along the solution trajectory as it crosses that line, and the maximum in U at the line of verticals naturally dominates over whatever smooth variation is occurring in the former derivative throughout this region.

10.B.3   *Forensic use of the UV plane: conclusions*

What are my conclusions about all this? The morbid fixation on equation (10.B.1), and in particular on the 'smallness' of $2U + V - 4 (= D)$, has led to an overemphasis on where the solution trajectories cross the line of tangents, or to misleading statements about such trajectories 'running along' that line, as though proximity to it per se were the major factor. That this is not so is indicated by comparing UV-plane trajectories for horizontal-branch models that have loops with those for red giants. The loops below the line of tangents stay much closer to it in the former case, and yet such stars have much smaller radial extents beyond their (radially larger) cores, ending up smaller in absolute terms for a given mass, core-mass and composition. This is predominantly because the range covered by the variable V/U during the loop is smaller, a feature more related to the understandable, overall topology of the UV plane. The latter pretty much follows from the value of U at the base of the envelope, though with significant sensitivity to the starting value for V, also. The key difference from red giants is of course the fact mentioned parenthetically: that their cores start out larger. Because of that, as our generalization of the density-seesaw theorem shows, their shell densities and therefore the values of U at the bases of the envelopes are larger.

The magnitude of U at the base of the envelope and the corresponding value of V are the major determinants of the topology of the remaining structure in the UV plane. Despite this significant fact, none of the UV-plane investigators ever seem to have bothered in their analyses to determine what U_{base} would actually be. Nor have they placed the right limit on the corresponding V_{base}, i.e. a value that would put the starting point below the line of horizontals, however much or little below. So, they lack a fundamental guide to the rest of the topology. And, as stated above, that topology shows that a given solution trajectory never 'runs along' the line of tangents. When it is running most 'parallel' below that line, it is of course furthest away from it. As one follows the solution along its trajectory beyond such a point, the departure from the line of tangents naturally decreases, but the value of $d \ln r/ds$ varies fairly smoothly, reaching a perfectly natural limit as it crosses the line. In passing along that given trajectory, there is nothing to demand that the dominant contribution to the full additional radial extent must peak at small separation D, and our examples confirm that the major contribution is often well beyond the line of tangents where D is becoming larger and U/V is now decreasing again towards its surface value. If one wishes to 'explain' the large size of the resulting models in terms of their properties in the UV plane, the explanation must have as much if not more to do with the slowly changing values of $d \ln r/ds$ being maintained throughout their long looping paths in that plane as it does with some claimed proximity of the line $D = 0$ per se for only part of those paths. But all these *post facto* UV-plane discussions are seriously deficient, and miss the point in a key respect: they merely *confirm* that the ultimate value of a particular star's total radius R is indeed large in terms of the path that has already been found to have been taken by its trajectory in the UV plane. The more fundamental question for UV-plane 'explanations' should surely be: '*Why* does the path have to be so long and so looping in the first place?' The answer to that is what I have shown above. The remaining envelope mass simply cannot be stored unless reasonably large values of U are accessed (values much larger than those explicitly demonstrated to hold for U_{sh} in our approach), and *that* can only be done with a path that loops out to such values, the only way of getting there, topologically. Thus the significance of U being large at some point in the envelope solution is not just that it weights mass storage to larger values of r (the almost parenthetical rôle assigned it by Fujimoto and Iben). The *need* for it to be large, to store the envelope mass *at all*, determines the global nature of the solution trajectories. The ultimately large value of R for the complete star simply comes along for the ride with this long and looping topological requirement, which is driven by mass-storage considerations. As I have shown, neither the satisfaction of envelope mass storage nor the ultimate value of R is in any way dominated by contributions arising from 'crossing' or 'running along' the line of tangents. All these claims or minute *post facto* examinations of the behaviour of already calculated models in the UV plane seem much like the study of stellar-model entrails, and from the point of view of making any explicit quantitative predictions they are no more useful than that. The fascination has become one with the tool itself, rather than with the stars to which it is applied.

References

APPLEGATE, J. H. 1988 *ApJ*, **329**, 803

BURBIDGE, E. M., BURBIDGE, G. R., FOWLER, W. A. & HOYLE, F. 1957 *Rev. Mod. Phys.*, **29**, 547

DORMAN, B. 1992 *ApJS*, **80**, 701

EDDINGTON, A. S. 1926, *The Internal Constitution of the Stars* (Cambridge: Cambridge University Press) p. 101

FAULKNER, J. 1964 *Computations in Stellar Structure* (Ph.D. Thesis, University of Cambridge)

    1966 *ApJ*, **144**, 978

    1967 *ApJ*, **147**, 617

    1976 in *Structure and Evolution of Close Binary Systems* (Proc. IAU Symp. No 73), eds. P.P. Eggleton, S. Mitton & J. Whelan (Dordrecht: Reidel) p. 193

    1997 in *Advances in Stellar Evolution*, eds. R.T. Rood & A. Renzini (Cambridge: Cambridge University Press) p. 9

FAULKNER, J. & IBEN, I., JR 1966, *ApJ*, **144**, 995

FAULKNER, J., EGGLETON, P.P., GILLILAND, R.L., & HOYLE, F. 1982 *BAAS*, **14**, 956

FUJIMOTO, M.Y. & IBEN, I. JR 1991 *ApJ*, **374**, 631

HASELGROVE, C.B. & HOYLE, F. 1956a *MNRAS*, **116**, 515

    1956b *MNRAS*, **116**, 527

    1958 *MNRAS*, **118**, 519

    1959 *MNRAS*, **119**, 112

HEN, LI & SCHWARZSCHILD, M. 1949 *MNRAS*, **109**, 631

HENYEY, L.G., WILETS, L., BÖHM, K.H., LELEVIER, R. & LEVEE, R.D.
    1959 *ApJ*, **129**, 628

HOYLE, F. 1946 *MNRAS*, **106**, 255

    1959 *MNRAS*, **119**, 124

    1994 *Home is Where the Wind Blows* (Mill Valley, CA: University Science)

HOYLE, F. & LYTTLETON, R.A. 1942a *MNRAS*, **102**, 177

    1942b *MNRAS*, **102**, 218

    1946 *MNRAS*, **106**, 525

    1949 *MNRAS*, **109**, 614

HOYLE, F. & SCHWARZSCHILD, M. 1955 *ApJS*, **2**, 1

IBEN, I. JR 1967 *ARA&A*, **5**, 571

KIPPENHAHN, R. & WEIGERT, A. 1990 *Stellar Structure and Evolution* (Berlin: Springer-Verlag) p. 321

ÖPIK, E. 1938 *Publications de l'Observatoire Astronomique de l'Université de Tartu*, 30/3, 1; 30/4, 1

OSTERBROCK, D.E. 1953 *ApJ*, **118**, 529

REFSDAL, S. & WEIGERT, A. 1970 *A&A*, **6**, 426

RENZINI, A., GREGGIO, L., RITOSSA, C. & FERRARIO, I. 1992 *ApJ*, **400**, 280

SANDAGE, A.R. & SCHWARZSCHILD, M. 1952 *ApJ*, **116**, 463

SCHÖNBERG, M. & CHANDRASEKHAR, S. 1942 *ApJ*, **96**, 161

SCHWARZSCHILD, M. 1958 *Structure and Evolution of the Stars* (Princeton: Princeton University Press) 91, 217, 260

SOUTHWELL, R. V. 1940 *Relaxation Methods in Engineering Science: A Treatise on Approximate Computation* (Oxford: Clarendon Press)

SUGIMOTO, D. 1997 in *Advances in Stellar Evolution*, eds. R. T. Rood & A. Renzini (Cambridge: Cambridge University Press) p. 19

SUGIMOTO, D. & FUJIMOTO, M. 2000 *ApJ*, **538**, 837

SUGIMOTO, D. & NOMOTO, K. 1980 *Space Sci. Rev.*, **25**, 155

WEISS, A. & SCHLATTL, H. 2000 *A&AS*, **144**, 487

WHITWORTH, A. P. 1989 *MNRAS*, **236**, 505

YAHIL, A. & VAN DEN HORN, L. 1985 *ApJ*, **296**, 554

11

Modern alchemy: Fred Hoyle and element building by neutron capture

E. MARGARET BURBIDGE

Center for Astrophysics and Space Sciences,
University of California at San Diego

Fred Hoyle's fundamental work on building the chemical elements by nuclear processes in stars at various stages in their lives began with the building of elements around iron in the very dense hot interiors of stars. Later, in the paper by Burbidge, Burbidge, Fowler and Hoyle, we four showed that Hoyle's 'equilibrium process' is one of eight processes required to make all of the isotopes of all the elements detected in the Sun and stars. Neutron capture reactions, which Fred had not considered in his epochal 1946 paper, but for which experimental data were just becoming available in 1957, are very important, in addition to the energy-generating reactions involving hydrogen, helium, carbon, nitrogen and oxygen, for building all of the elements. They are now providing clues to the late stages of stellar evolution and the earliest history of our Galaxy. I describe here our earliest observational work on neutron capture processes in evolved stars, some new work on stars showing the results of the neutron capture reactions, and data relating to processes ending in the production of lead, and I discuss where this fits into the history of stars in our own Galaxy.

11.1   Introduction: the origin of B^2FH

I was a post-doc at the University of London Observatory, Mill Hill, in 1946. The Director, C. C. L. Gregory, was accustomed to escorting the junior people at the Observatory to the monthly meetings of the Royal Astronomical

Society in Burlington House, Piccadilly, London. Thus I was one of the group sitting in the auditorium when Fred Hoyle read his famous paper (Hoyle 1946): 'The synthesis of the elements from hydrogen'.

This was at a time when George Gamow, Maria Goeppert-Mayer and Edward Teller were in full spate with the theory that all the chemical elements were created primordially by the coagulation of neutrons just after the birth of the Universe (see Mayer and Teller 1949). Fred's paper was presented on 8 November 1946, with Professor H. H. Plaskett, President of the Royal Astronomical Society, in the chair.

Listening to Fred's presentation, I sat in the RAS auditorium in wonder, experiencing that marvellous feeling of the lifting of a veil of ignorance as a bright light illuminates a great discovery.

While I was not privileged for seven more years to become a colleague of Fred's and to work with him on stellar nucleosynthesis, Fred's 1946 work began the move of my scientific interest from Ph.D. work on a class of hot stars to the challenge of determining by observation the chemical composition in the atmospheres of some unusual stars. Work by Lawrence Aller, Joseph Chamberlain, Nancy Roman and others had established the low metal abundances of Population II stars, and the spectra of some A-type stars with strong and variable magnetic fields showed large apparent and variable overabundances of certain chemical elements on their surfaces, which invited investigation.

My husband Geoffrey and I were at the Yerkes Observatory, University of Chicago, in 1951–3, and were assigned time on the 82-inch telescope at McDonald Observatory in SW Texas in the spring of 1953. We obtained high-resolution spectra of the A-type magnetic variable α^2 CVn, and worked day and night to run microphotometer tracings of these in the Yerkes basement, before leaving the USA for Cambridge, England, where Geoff had been appointed to a position in Martin Ryle's radio astronomy group in the old Cavendish Laboratory.

Curve-of-growth analysis of the α^2 CVn data, which we worked on in Cambridge, led to overabundances of a number of chemical elements, particularly certain heavy elements, which suggested to us that somehow neutrons were involved – an idea whose germ had been planted by the Mayer–Teller and Gamow work, but we believed the nuclear processes must take place in stars.

Geoff gave a scientific talk on this work and on our attempt to understand the observed overabundances of the rare earths and some 'magic number' (closed shell) nuclei by nuclear processes involving the acceleration of protons on the surfaces of magnetic stars. After the talk a very cheerful physicist came up to Geoff and told him that he was an experimental nuclear physicist visiting Cambridge as a Fulbright Professor from Caltech. This was William A. (Willy) Fowler. He was intrigued by our results, but told Geoff that he only worked on the light elements! He pointed out that Fred Hoyle, now well known for his work on building the iron-peak elements in stars, lived in Cambridge, and we four could get together and work on the nuclear physics going on in stellar interiors. He introduced us to Fred Hoyle, and we began working together.

By the end of summer 1955 we had outlined eight processes on which we should work: hydrogen burning, helium burning, the alpha process, Fred's e-process updated, two neutron-capture processes (slow-s, and rapid-r), a proton (p) capture process to account for some less-abundant heavy isotopes in solar and terrestrial data, and what we later called the x-process, because it was the least understood, to account for the elements lighter than carbon. At the end of summer 1955, Fowler left Cambridge and returned to Caltech. Geoff and I left Cambridge for Pasadena, where Geoff had been awarded a Carnegie Fellowship and I, a post-doctoral fellowship at the Kellogg Radiation Laboratory, Caltech. Fred came to Pasadena in early 1956.

11.2   Synthesis of the elements in stars

This was the title of our completed work, published in 1957. We headed it with two captions from Shakespeare:

> It is the stars,
>> The stars above us, govern our conditions (King Lear, Act IV, Scene 3)

and,

> The fault, dear Brutus, is not in our stars,
>> But in ourselves (Julius Caesar, Act I, Scene 2).

These are still apposite. No one now denies that the chemical elements of which all of us and everything around us are composed were built from hydrogen by nuclear reactions in generations of stars, but new observations with telescopes and equipment hardly dreamed of 45 years ago are continually building on earlier results, and the story is still unfolding.

An account of progress during the 40 years after B^2FH was published by George Wallerstein (1997). He assembled discussions by 14 experts on the state of the art in stellar evolution and in each of the eight processes of B^2FH, plus an immense list of references; the result, 'Synthesis of the elements in stars: forty years of progress', was published in the same journal as B^2FH, the *Reviews of Modern Physics*, and is about the same length as B^2FH.

The schematic diagram showing abundances of the elements in the Solar System plotted against atomic weight A, based on work by Suess and Urey (1957), was Figure 1 of B^2FH. Besides indicating the elements involved in the

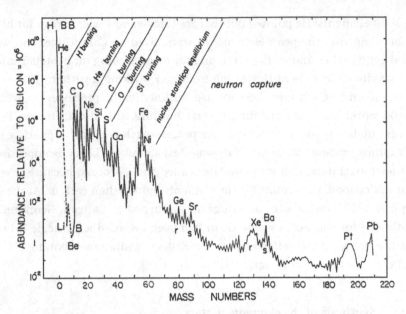

Figure 11.1 Diagram showing relative abundances of elements in the Solar System, Figure 1.4 in Pagel (1997).

energy-releasing nuclear reactions, up to the 'iron group', it showed clearly the double peaks marked r and s in elements heavier than the iron-peak elements, at places where nuclei have closed shells of neutrons, which we indicated as $N = 50$, 82 and 126. Numerous versions of these data have appeared in textbooks and elsewhere; a good one is shown in Figure 11.1, reproduced from Pagel (1997).

In the slow neutron capture (s-process), elements in the iron abundance peak capture neutrons if these are produced within an evolved star. The reactions proceed until nuclei with closed shells of neutrons are built, at which points the capture cross sections drop abruptly, and peaks in abundance are produced at these points. This followed Maria Goeppert-Mayer's shell model of nuclei; our analogy in B^{2}FH was of water flowing over a river bed and encountering deep holes in the bed, where water would accumulate until the hole was filled, after which time it would flow on to the next hole. Our relevant timescale between neutron captures was ~10^5 years.

In the rapid neutron-capture process a large flow of neutrons passes over the s-process 'holes', and drives the nuclear matter far to the neutron-rich side of the stability line, until the neutron binding energy decreases to 2 MeV – our estimate in B^{2}FH – at which point beta decay must occur before another neutron can be added, thus producing the displaced abundance peaks (Figure V, 2 in B^{2}FH and numerous textbook figures show these capture paths).

Fred Hoyle's 1946 paper showed that stars in which evolution had gone as far as the e-process would be completely unstable. Hence, stellar collapse and explosion as a supernova must follow. Release of a flood of neutrons, we thought, would lead to captures following our outline of the r-process. Dispersal of the enriched material into the surrounding interstellar medium by supernova explosions would make this enriched material available for subsequent star formation. Europium would be a good indicator of material that had been thus enriched, as it was a characteristic product of the r-process. Meyer *et al.* (1992) have suggested that the r-process occurs in the hot high-entropy bubble surrounding the nascent neutron star during the supernova explosion. The r-process calculations (see Cowan *et al.* 1999) are based on $(n, \gamma) \rightleftarrows (\gamma, n)$ equilibria, which will be obtained in astrophysical environments where $n_n \geq 10^{20}\,\mathrm{cm}^{-3}$ and $T \sim 10^9$ K.

11.3   Slow neutron capture: early observational data

Calculations of late stages of stellar evolution now make it possible to model the mixing and dredge-up processes occurring in stars on the asymptotic red-giant branch of colour–magnitude diagrams. Spectrographs and modern detectors make it possible to determine element abundances from high-dispersion spectra of very metal-poor stars. The history of stars with overabundances of s-process elements, their occurrence as members of close binaries, etc. is reasonably well understood now. In 1955, however, the only indication that the s-process could occur at a late evolutionary stage was the detection of the unstable element technetium in red-giant stars by Merrill (1952).

During the work on B²FH in Pasadena, Fred, Geoff and I worked at Caltech in one long windowless room in the Kellogg Radiation Laboratory building round the corner from Willy Fowler's office. Knowledge of stellar evolution beyond the move from the main sequence into red-giant structure was at an early stage in 1957. But it was obvious that a helium core would build up, and the existence of carbon-rich stars demonstrated that Salpeter's (1952) triple-alpha process occurred. Some evolved stars showed ^{13}C as well as ^{12}C, indicating that the ^{13}C$(\alpha, n)^{16}$O reaction could occur and produce neutrons in red-giant stars, and, we thought, possibly ^{22}Ne$(\alpha, n)^{25}$Mg as well. Without being able in 1957 to follow stars onto the asymptotic red-giant region, let alone into stellar collapse and stellar explosion, our general picture was that spurts of mixing in giant stars between a carbon-rich inner region and a helium-rich intermediate region could produce neutrons which, captured by already existing iron-peak elements, could result in stars showing overabundances of the s-process elements such as strontium, yttrium, zirconium in the first s-process peak, barium in the second, and lead in the third.

As we worked on the s-process in Pasadena we realized that we needed observational data on an evolved star that showed overabundances in its atmosphere of heavy elements produced in this process, i.e. stars of spectral class S, and that class of G and K giant stars named 'Ba II stars' by Bidelman and Keenan (1951) because of the great strength of Ba II λ4554 in their spectra. The brightest Ba II star, ζ Cap, had been observed by J. L. Greenstein, and his spectra, awaiting analysis, were not available to us. Geoff Burbidge, as a Carnegie Fellow, was entitled to apply for observing time on the Mt Wilson telescopes, but I was the observational member of our partnership, and women were not allowed on Mt Wilson; Director Ira Bowen was in agreement with that ancient policy.

Our friend Allan Sandage, at 813 Santa Barbara Street, had begun regular discussions with Bowen on several matters of Mt Wilson Observatory policy, and the ban on allowing women astronomers to use its telescopes was one. Caltech was also joining this effort, armed with information from Willy Fowler, who was pointing out that such discrimination was no longer tolerable.

With reluctance, I suspect, Director Bowen withdrew his opposition; Geoff was allocated observing time on the 100-inch and 60-inch telescopes, and I was allowed to accompany him as long as I kept a low profile: we should use our own transport up the mountain, stay in the small summer cottage (the Kapteyn Cottage) rather than the dormitory and dining room where astronomers normally stayed, and bring our own food.

For a star showing surface abundances due to the s-process, we chose HD46407, the next brightest accessible star (after Greenstein's ζ Cap) from the list of Ba II stars in the important paper by Bidelman and Keenan (1951). For comparison, we chose κ Gem (G8 III) as the standard star, and obtained Coudé-grating spectra of both stars with the 100-inch telescope in November 1955 and January and February 1956. Measurement, data reduction, and analysis were carried out at both Kellogg and 813 Santa Barbara Street, and the results – wavelengths, identifications, and equivalent widths of the spectrum lines – were published (Burbidge and Burbidge 1957). I reproduce in Figure 11.2 our figure showing portions of our spectra of HD46407 and κ Gem. This is the way in which we reproduced photographic spectra 45 years ago!

I recall Fred and Geoff sitting in that Kellogg office, each operating one of those 1940–50-era calculating machines, grinding away at the calculations for the r-process and the s-process, while I sat in the same room, ploughing away through the data on HD46407. We obtained abundances of a number of s-process elements, in comparison with the same elements in the standard star, κ Gem, which had the same atmospheric conditions. These elements would contain some isotopes not produced in the s-process, so using the nuclear physics analysis that was emerging from our work to correct the overabundances, y, for those isotopes produced by the r- and the p-processes, thus obtaining y' for the s-isotopes alone, we multiplied y' by the neutron capture cross section σ for each relevant element. The product σN, N being the solar (and κ Gem) abundance

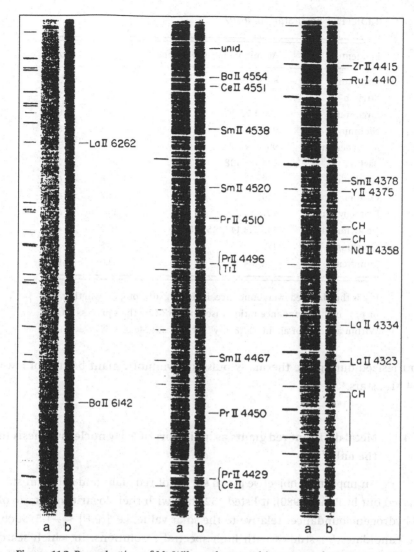

Figure 11.2 Reproduction of Mt Wilson photographic spectra of HD46407 and κ
Gem, Figure 1 in Burbidge and Burbidge (1957).

times the observed overabundance in HD46407, should be constant. Although
the cross sections available to us were not very accurate (they were improved
substantially soon afterwards), this product for 12 s-process elements was, within
a factor of approximately 2, constant (Burbidge and Burbidge 1957). Our data are
reproduced in Table 11.1. An account of more modern work on the s-process is
given in Section XI, by Verne V. Smith, in the 'Forty years of progress' article by
George Wallerstein (1997). The lighter s-process elements are probably produced
in advanced evolutionary phases of massive stars, while the heavier (Sr to Pb) are

Table 11.1 *Ba II Star HD46407*

Element	Atomic Weights, s-Isotopes	y'	$\langle \sigma N \rangle$
Strontium	86,87,88	4.7	381
Yttrium	89	7.8	1326
Zirconium	90,91,92,94	4.9	884
Niobium	93	5.2	832
Molybdenum	95,96,97	4.8	343
Barium	134,136,138	14.9	281
Lanthanum	139	9.8	647
Cerium	140	10.7	407
Praseodymium	141	28	672
Neodymium	142,143,144,145,146	15.7	443
Ytterbium	170	2.8	78
Tungsten	182,184	13.0	914

If y is the observed overabundance, including all isotopes, whether s, r or p, y' is the abundance ratio of isotopes built by the s-process alone. Within a factor of about 2, $\langle \sigma N \rangle y'$ is constant: Mean $\langle \sigma N \rangle y' = 648$.

synthesized during the thermally pulsing asymptotic giant branch of low-mass, 1–4 $M_\odot$, stars.

11.4   Metal-deficient red giants as indicators of early nucleosynthesis in the milky way

An important objective-prism survey of red-giant and subgiant stars was carried out by Bond (1980); it listed 132 stars with their logarithmic ratio of iron to hydrogen abundances relative to the solar value, i.e. [Fe/H] $\leq$ −1.5. Such stars are candidates for studies with high spectral resolutions, in which search for and measurement of the spectral line intensities can be made for rare elements produced in the neutron-capture chains. Stars with the lowest [Fe/H], i.e. [Fe/H] $\leq$ −3.0, do not appear to show the products of the s-process in their spectra, and this is important information for elucidating the timescale for formation and evolution of stars early in the history of the Galaxy.

Sneden *et al.* (1996) obtained data at the Cerro Tololo Inter-American Observatory for the very metal-poor star CS22892-052, a K giant star (from the survey by Beers *et al.* (1992)). This star has [Fe/H] $\simeq$ −3.1, so is ultra-metal-poor. Sneden *et al.* derived abundances for 20 heavy elements from strontium to thorium. Figure 11.3 reproduces their Figure 4, showing the high relative abundances of europium and other r-process elements. Of particular interest are their

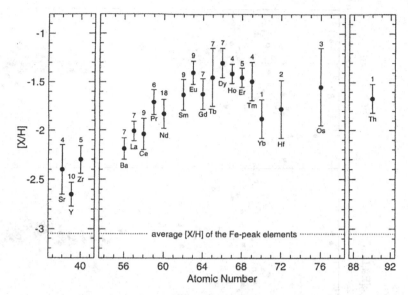

Figure 11.3 Reproduction of Figure 4 from Sneden *et al.* (1996), showing the high relative abundances of Eu and other r-process elements in the ultra-metal-poor giant star CS22892-052.

detections and measurement of holmium and terbium. Holmium had been detected in the peculiar star HD101065 (Przybylski 1961), but the terbium detection was a new result.

They also detected and measured the radioactive element thorium, a pure r-process product. It was detectable by only one available line, at 4019 Å, but the identification appears secure. Since thorium is radioactive (with a half-life 14.0 Gyr), its detection enables an estimate to be made for the age of this ultra-metal-poor star; Sneden *et al.* derived 15.2 ± 3.7 Gyr. A study of r-process abundances as chronometers in metal-poor stars by Cowan *et al.* (1999), using the [Th/Eu] ratio in the very metal-poor halo stars HD115444, CS22892-052 and HD122563, produced a similar age estimate: 15.6 ± 4.6 Gyr.

Further work on n-capture elements in very metal-poor stars was carried out with Hubble Space Telescope data by Sneden *et al.* (1998), and by Burris *et al.* (2000) using the KPNO 4-m telescope, on a large sample of metal-poor giants from the list of Bond (1980).

Sneden, in Chapter XIII of Wallerstein *et al.* (1997), plotted [heavy-element/Fe] logarithmic abundances versus [Fe/H] for a number of elements produced in the s- and r-processes. I reproduce in Figure 11.4 his Figure 21, where the top figure is for the 'light' s-process elements, the middle, for the 'heavy' s-process elements, and the bottom, the r-process elements; all are plotted against [Fe/H]. The difference between these plots for elements characteristic of the r- and

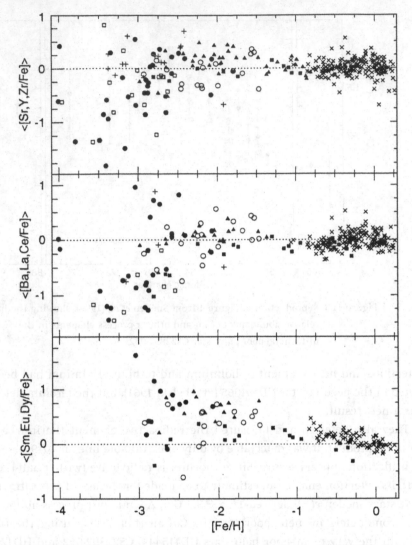

Figure 11.4 Reproduction of Figure 21 from Sneden's Chapter XIII in Wallerstein
et al. (1997), showing the trends of the 'light' s-process elements (top
panel), 'heavy' s-process elements (middle panel), and r-process elements
(bottom panel), with decreasing stellar metallicity [Fe/H]. The relative
abundances are defined as $[A/B] \equiv \log_{10}(N_A/N_B)_{star} - \log_{10}(N_A/N_B)_?$ for
elements A and B.

s-processes is striking. From [Fe/H] $\sim$ 0 to -1, there is little scatter about the
dotted line, which represents the solar value, but for [Fe/H] < -1 the scatter
increases for both the s-process and r-process products, but the trend towards
decreasing [Fe/H] is different for the two neutron capture chains. The upper two
diagrams show a downward trend (with much scatter) for the s-process, while
the bottom figure, for the r-process, trends upwards.

In their paper Burris *et al.* (2000) describe the sequence of heavy-element en-richment as follows: the initial production of Ba is most likely to be the r-process alone, occurring in supernovae resulting from more massive, faster-evolving stars. The first material of s-process origin occurs at [Fe/H] as low as -2.75, but the gradual transition to s-process occurs only at [Fe/H] $= -2.4$. Several studies, as referenced by Burris *et al.* (2000), suggest that most r-process elements are produced in Type II supernovae occurring in the explosion of stars in the 8 to 10 $M_\odot$ mass range.

11.5   The element lead: end of the s-process

The element lead is the last stable element produced in the slow neutron capture chain of nucleosynthesis. Decades ago we noted that Struve and Swings thought they might have detected lines of PbI and PbII in the magnetic star α^2 CVn. In our paper on that star, we examined the data and published our results in Table 7 of that paper. We did not detect lines at wavelengths near PbI 3639.57 and 3683.47, whereas Struve and Swings detected lines but left them unidentified. However, we detected lines near two PbII lines with λ(laboratory) $=$ 4245.1, 4386.4, for which we could provide no other identification. Thus, we concluded that Pb was present in α^2 CVn, and that its abundance was increased over solar by a factor of the same order as that for the rare earths. This result has never been challenged, and deserves to be investigated with modern data.

An exciting result recently published in the European Southern Observatory's *Messenger* (Van Eck *et al.* 2001) gives data obtained by this group with the 3.6-m telescope and the very long camera of the CES ultraviolet echelle spectrograph of the European Southern Observatory in Chile (Figure 11.5). Using an instrumen-tal setup giving a resolution R $=$ 135 000, they observed a group of CH (carbon-rich) evolved stars, HD196944, HD187861, HD224959, HD218875. Between wave-lengths 4057.4 and 4058.3 Å, their spectra clearly showed the line PbI 4057.81 in HD196944, HD187861, and HD224959, and showed that it was absent in the otherwise similar star HD218875. This is an important observation, which made use of the very high spectral resolution available at ESO, and will enable further study of the evolutionary path and nucleosynthetic history of stars in the CH classification.

11.6   Conclusion

The products of the r-process, seen in ultra-metal-poor stars in the halo of the Milky Way, show that supernovae as end stages of massive stars early in the history of the Milky Way seeded the halo of the still-forming Galaxy with these heavy elements and provide a clock for the age of formation of the halo.

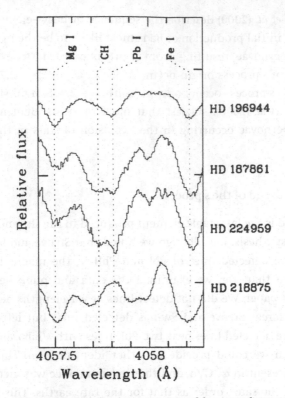

Figure 11.5 High-resolution spectra of four stars between 4057.5 and 4058.5 Å, obtained by Van Eck *et al.* (2001) with the 3.6-m telescope at the European Southern Observatory. The PbI line is clearly seen in the CH stars HD196944, HD187861, and HD224959, and is as clearly absent in the R-type comparison star HD218875. Figure taken from Van Eck *et al.* (2001).

The most metal-poor of the globular clusters similarly provide data on their age since formation. A review by Vandenberg *et al.* (1996) on the age of the Galactic globular cluster system gives 15 Gyr for this age. A study of three red giants at the tip of the red-giant branch in the globular cluster M15 (Sneden *et al.* 2000) has shown that the abundances of the heaviest r-process elements yield a similar age: the weighted mean observed ratio of thorium to europium in M15 gives an age 14 ± 3 Gyr.

The data of Burris *et al.* (2000) make it clear that the slow neutron capture s-process sets in later in the history of the Galaxy. I believe the main contribution of stars showing results of the s-process to our understanding of the history of the Galaxy will come from the light they throw on the complex late-evolutionary stages of stars, and the several 'dredge-up' episodes of mixing between layers, which at present challenge the best computers and computational experts.

We are at an exciting stage in astrophysics: new large telescopes will soon be joining, and surpassing, the 10-metre-class telescopes, so fainter stars will become accessible, and advances in instrumentation and the steadily advancing capability of computers are together promising exciting results in the details of stellar evolution.

Fred Hoyle would have been delighted to make sense of the new data – the symposium gave so much evidence of how sorely we miss him!

References

BEERS, T. C., PRESTON, G. W. & SHECTMAN, S. A. 1992 *AJ*, **103**, 1987

BIDELMAN, W. P. & KEENAN, P. C. 1951 *ApJ*, **114**, 472

BOND, H. E. 1980 *ApJS*, **44**, 517

BURBIDGE, E. M. & BURBIDGE, G. R. 1957 *ApJ*, **126**, 357

BURBIDGE, E. M., BURBIDGE, G. R., FOWLER, W. A. & HOYLE, F. 1957 *Rev. Mod. Phys.*, **29**, 547

BURRIS, D. L., PILACHOWSKI, C. A., ARMANDROFF, T. E., SNEDEN, C., COWAN, J. J. & ROE, H. 2000 *ApJ*, **544**, 302

COWAN, J. J., PFEIFFER, B., KRATZ, K.-L. *et al.* 1999 *ApJ*, **521**, 194

HOYLE, F. 1946 *MNRAS*, **106**, 343

MAYER, M. G. & TELLER, E. 1949 *Phys. Rev.*, **76**, 1226

MERRILL, P. W. 1952 *ApJ*, **116**, 21

MEYER, B. S., MATHEWS, G. J., HOWARD, W. M., WOOSELY, S. E., HOFFMAN, R. D. 1992, *ApJ*, **399**, 656

PAGEL, B. E. J. 1997 *Nucleosynthesis and Chemical Evolution of Galaxies* (Cambridge: Cambridge University Press) p. 8

PRZYBYLSKI, A. 1961 *Nature*, **189**, 739

SALPETER, E. E. 1952 *ApJ*, **115**, 326

SNEDEN, C., McWILLIAM, A., PRESTON, G. W., COWAN, J. J., BURRIS, D. L. & ARMOSKY, B. J. 1996 *ApJ*, **467**, 819

SNEDEN, C., COWAN, J. J., BURRIS, D. L. & TRURAN, J. W. 1998 *ApJ*, **496**, 235

SNEDEN, C., JOHNSON, J., KRAFT, R. P. *et al.* 2000 *ApJL*, **536**, L85

SUESS, H. E. & UREY, H. C. 1957 *Rev. Mod. Phys.*, **28**, 53

VANDENBERG, D. A., BOLTE, M. & STETSON, P. B. 1996 *ARA&A*, **34**, 461

VAN ECK, S., GORIELY, S., JORISSEN, A. & PLEZ, B. 2001 *The Messenger*, December

WALLERSTEIN, G. *et al.* 1997 *Rev. Mod. Phys.*, **69**, 995

12

Concluding remarks

GEOFFREY BURBIDGE

Center for Astrophysics and Space Sciences,
University of California at San Diego

This final chapter is based on notes prepared for an after-dinner speech at the dinner ending the day of celebration of Fred Hoyle and his science in Cambridge, 16 April 2002.

We have celebrated Fred Hoyle's scientific contributions in many areas, and nearly all of the major themes have been touched on. In making a final toast to Fred, let me recount a little about my own relation and work with him. I think that it is also proper to say something about aspects of his achievements that have not been stressed enough or even mentioned on the day.

I started to work with him 47 years ago in Cambridge where I was officially working in the Cavendish with the radio astronomy group. My wife Margaret and I met Willy Fowler, who was a visiting Fulbright Professor, and began work on slow neutron capture processes in stars, which was later named the s-process in B²FH. We met Fred through Willy, who had already got to know him when Fred visited Caltech before 1954. Through 1954 and 1955 we began to work intensively together in Cambridge, and in the autumn of 1955 three of us went to California permanently, and Fred made a number of very extensive visits to Caltech. Our work was written up and published in 1957. This was one of the most exciting periods that I have ever experienced. Fred was the founding genius of nucleosynthesis, but we were all able to bring different kinds of expertise to the endeavour. In this period I was also working on the energetics of radio sources, and began to try to understand them with Fred and the others. This was the beginning (for me) of Fred's interest in high-energy astrophysics, and it

The Scientific Legacy of Fred Hoyle, ed. D. Gough.
Published by Cambridge University Press. © Cambridge University Press 2004.

led in the late 1950s and early 1960s to the idea that energy was released from chains of supernovae (my idea which Fred and Willy didn't like) and then to the collapse of supermassive stars. By then, Margaret and I were firmly ensconced in the United States, and joint projects were carried out on Fred's frequent visits to Pasadena and La Jolla and our summer visits to Cambridge.

We were in Cambridge in the planning stages of the creation and building of IOTA and through all of the summers that followed. What I would like to stress is just how hard that project was. When Fred first proposed that such an institute be created, the proposal was supported on the highest level by the appropriate national committees including the Astronomers Royal, but when it was formally proposed to the University of Cambridge, a committee was set up by Sir Nevill Mott, at that time the Cavendish Professor, and they turned it down. Fred was extremely upset, and tried to resign his chair. He was persuaded not to do this by people like Sir Alan Cottrell and Dame Mary Cartwright. Several real heavyweights strongly supported him, in particular, Alex Todd (Lord Todd) and Sir John Cockcroft. They forced a reversal of the university's earlier position, and also arranged funding through the Nuffield Foundation, the Wolfson Foundation, and the Science Research Council (SRC). The land for the building was to come from the University of Cambridge. The final snag developed when the SRC said that they were prepared to fund the Institute, and that it could be located anywhere, *provided it was not in Cambridge!* Of course, this last piece of nonsense, which was to do with the governance of the Institute, was reversed, but not before Fred again talked of taking the whole thing to Inverness. But as you know, it went through and was a great success, which has continued even after the upheaval in 1972 which ended with Fred's departure. This second battle Fred lost. This second battle was even worse than the first one to get IOTA. Many of the same cast of characters who had originally been opposed came out of the woodwork, and this time, by devious methods, they forced a situation on Fred which led to his resignation. What I took away from this was how incredibly hard it is to do something new and innovative in such an environment.

Of course, in staffing the Institute, Fred very quickly put together a first-class group of people, and let me make another point here. Fred believed in very good people, creative individuals, and he paid little or no attention to what opinions they might hold, nor whether or not they might agree with him. In this he was very different from many other leaders of that time, and now. Particularly in cosmology, conformity was the order of the day (then and now). Bob Dicke in Princeton, Ya B. Zeldovich in Moscow, and Martin Ryle in Cambridge, never had anyone working with them who did not slavishly follow their leader. Of course, I think that Fred's method is far superior, but if your goal is to win men's minds (as they say in the USA), i.e. to win medals, get funds and, above all, convince people that your view is correct, obviously the others have done better. But Fred always felt that what we are trying to do is to understand a very complex universe rather than to conform to a set of ideas already believed in by great men not necessarily for good reasons.

The last thing I want to talk about is the balance of Fred's research. You have read good accounts of his forays into accretion, stellar structure and evolution, cosmogony, nucleosynthesis, interstellar matter, dust, and organic molecules, and some discussions of cosmology, particularly in an excellent chapter by Jayant Narlikar. But not enough has been said concerning the highly creative work that he continued to do from the 1970s until his death. His chief collaborators in this latter period were Chandra Wickramasinghe, Jayant Narlikar and me. As always, Fred was attempting to understand many of the new observational discoveries. It is not surprising that little has been said of much of the work done in this period, because it is still not accepted or believed by many. How long it will take for this to change I do not know, and in some cases, it may turn out that we were wrong. But the fact that Fred's view was often a minority view probably means that he was far ahead of his time. (My friend V. L. Ginzburg always says, somewhat ruefully: 'The minority usually turns out to be right'.)

Take the quasi-stellar objects (QSOs). They were discovered in 1960, and by 1966 Fred (and I) began to believe that the observational evidence pointed to a local origin for many of them, ejected from comparatively nearby active galaxies. Despite the overwhelming desire of the community to believe that the redshifts of the QSOs must be entirely cosmological in origin, the evidence has continued to accumulate, most recently from X-ray observations, that many QSOs have large redshift components of non-cosmological origin. In the last decade or more, Fred became totally convinced that the nature of non-cosmological redshifts was one of the great unsolved problems of physics. He (and I) tried to find solutions, but have so far been unsuccessful. But the evidence for non-cosmological redshifts for QSOs is incontrovertible.

The other major issue is the cosmological problem. In the last book *A Different Approach to Cosmology* with Fred, published in 2000, the evidence for the quasi-steady-state cosmology (QSSC) was expounded in detail. In this book and in papers published since 1990, we found it possible to take a large amount of cosmogonical and cosmological evidence and show that it points very directly to a quasi-steady-state universe. The idea is that galaxies are formed and evolve through the ejection of matter, which takes place in nuclei of existing galaxies. The Big-Bang believers ignore all of the evidence concerning ejection from active galaxies. We believe that there was no big bang, but a continuous succession of little bangs. The arguments in favour of this model are powerful, and the evidence is strong. Jayant Narlikar was able to mention it briefly, but, like other unpopular ideas espoused by Fred, in recent years, and at the meeting, there has been a tendency to soft pedal it or ignore it altogether. In fact, it is quite frightening to see how far the conformists will go in ignoring different ideas. For example, the most recent discovery of the acceleration of the expansion of the Universe fulfils one of the major predictions of the classical steady-state cosmology of 1948, and of the QSSC, but the observers themselves refused to say so when they first discovered it; instead we are told the Big Bang is still okay, now with a finite cosmological constant, dark energy, quintessence, and what have you. Actually, it's creation (but don't say so).

In 60 years of research, Fred had a huge number of ideas. Like all very great scientists he was not afraid to be wrong. In my view, he was much more often right than wrong, because he was always guided by the evidence (all of it), and by an excellent intuition. In the latter part of his career his views diverged a long way from what most people want to believe, particularly in cosmology and high-energy astrophysics. Much of the difficulty here, in my view, has been due to the fact that ideas about origins and cosmology, in general, are being driven more today by sociology than by scientific evidence. Eventually we (or some of us) will know the answer. As a betting man, I will wager that again, in this second round of the battle between Galileo and the Vatican, Galileo (in the person of Fred) will win.

Enough. Finally let us drink a toast to a great man whose achievements have lifted us all: to Fred Hoyle.

References

BURBIDGE, E. M., BURBIDGE, G. R., FOWLER, W. A. & HOYLE, F. 1957 *Rev. Mod. Phys.*, **29**, 547

HOYLE, F., BURBIDGE, G. & NARLIKAR, J. V. 2000 *A Different Approach to Cosmology* (Cambridge: Cambridge University Press)

Index

Printed in the United States
By Bookmasters

Printed in the United States
By Bookmasters